PLACING
AUTOBIOGRAPHY
IN GEOGRAPHY

Space, Place, and Society
John Rennie Short, *Series Editor*

PLACING
AUTOBIOGRAPHY
IN
GEOGRAPHY

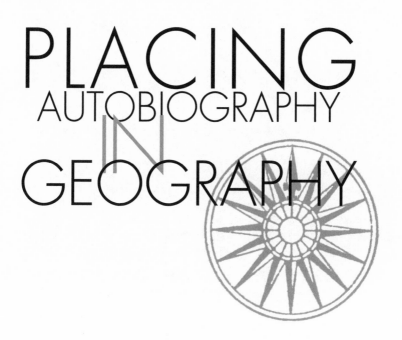

EDITED BY
PAMELA MOSS

Syracuse University Press

Copyright © 2001 by Syracuse University Press
Syracuse, New York 13244–5160

All Rights Reserved

First Edition 2001
01 02 03 04 05 06   7 6 5 4 3 2 1

The paper used in this publication meets the minimum requirements of
American National Standard for Information Sciences—Permanence of
Paper for Printed Library Material, ANSI Z39.48–1984.∞™

Library of Congress Cataloging-in-Publication Data
Placing autobiography in geography / edited by Pamela Moss.
    p. cm.—(Space, place, and society)
    Includes bibliographical references (p.) and index.
    ISBN 0-8156-2847-1 (alk. paper)—ISBN 0-8156-2848-X (pbk. : alk. paper)
        1. Geographers—Biography. 2. Geography—Study and teaching
        (Higher) 3. Autobiography—Study and teaching (Higher)
            I. Moss, Pamela J. (Pamela Jane) II. Series
    G67.P63 2000
    910'.922—dc21   00-038780

Manufactured in the United States of America

For Samuel and Hannah,
knowing they're writing their own lives

Contents

Illustration

Acknowledgments

Autobiography has a reputation of being difficult to write academically. I wonder sometimes if it isn't the writing that is difficult, but the reading: reading autobiography so that it makes sense in a context other than the one within which it is written, within which it exists.

As for me, my context includes Ann, Elizabeth, and Joan, all of whom have made me want to pursue this project by setting and living their lives in multiple contexts. Karl, too, has made me want to write and read autobiography in such a way that it is intelligible to the people closest to me. Clarice, my mother, has, almost inevitably, shaped who I am in ways I'll never be able to sort through.

This has been a solitary project and somewhat lonesome. But I had tremendous support from people around me, most of whom knew nothing of this book: the students in the 1997 seminar in "Spatial and Sexual Politics," students in the 1998 course in "Women in the City," and the graduate students and research assistants interested in autobiography that I've had the privilege of working with at the University of Victoria: Amy Zidulka, Andrea Lloyd, Geoff Whitehall, Jay Petrunia, Kathleen Gabelmann, Markus Navikenas, and Wanda Ollis. Of those who knew about the project, John Short, editor of the "Space, Place, and Society" series at Syracuse University Press, in particular, has been supportive of the idea from the beginning. J. P. Jones and Derek Gregory assisted me in identifying some potential contributors and encouraged me to pursue the project, particularly in the initial stages. I thank the contributors for their patience—most chapters were drafted by May 1998. Editors at Syracuse University Press made the publication process relatively painless! An anonymous reviewer made several suggestions to improve the quality of the collection for which I'm also grateful. I compiled the manuscript and completed the editing while a Visiting Professor in Human Geography at the University of Vienna.

Finally, I want to thank Suzanne Mackenzie for her advice, mentoring, and sense of humor. She made me think about my life in the context of others. I feel her death deeply.

Vienna, Austria   Pamela Moss
Summer 1999

Contributors

Kevin Archer teaches urban/social geography and the critical history of Western civilization in the Learning Community undergraduate program at the University of South Florida. His spouse, Dr. Ingrid Bartsch, teaches an innovative sequence of courses focused on the participation and role of women in science/technology fields for the Department of Womens' Studies and the program in Environmental Science and Policy. Archer received his Ph.D. from the Johns Hopkins University and his M.A. and B.A. from McGill University. His specific research interests include Disneyesque postindustrial city development and the meaning of "community," both inside and outside the academy.

Anne Buttimer is professor and chair of geography at University College Dublin. Graduate of University College Cork, she received her Ph.D. in geography at the University of Washington (Seattle) in 1965 and since then has held research and teaching positions in Belgium, Canada, France, Scotland, Sweden, and the United States. She is author of ten books and over a hundred articles on subjects ranging from social space and urban planning to the history of ideas and environmental policy. Some of her work has been published in translation in Dutch, French, German, Japanese, Portuguese, Russian, Spanish, Swedish, and Russian. She has received many awards and honors, among them the Ellen Churchill Semple Award and the Association of American Geographers' Honors Award. With Torsten Hägerstrand (Sweden) she initiated an international dialogue project (1978–88), directed Swedish-Canadian research on the human use of woodlands (1989–91) and coordinated an EU research network on landscape, life, and sustainable development with teams in Germany, Ireland, the Netherlands, and Sweden (1992–95).

David Butz is associate professor of geography at Brock University, St. Catharines, Ontario, Canada. His interests in social and cultural geography, geographies of everyday resistance, community level social organization in northern Pakistan and transport labor in the Karakorum/Himalaya coalesce in his current research project which investigates how portering relations have significantly shaped transcultural (insider/outsider) interactions in the Karakorum region. These empirical interests provide opportunities to build theory, which seeks to explain the geography of contemporary transcultural relations and transcultural discourses in the contact zone. Residents of Shimshal village, northern Pakistan, have significantly affected his autobiography and continue to shape his life.

Ian Cook is a lecturer in human geography at the University of Birmingham, U.K. He wrote his chapter in this book when he had the same job at the University of Wales, Lampeter. And he may or may not have got his Ph.D. from the University of Bristol, U.K. That's discussed in the chapter. At work, he claims to be a cultural/economic geographer. He knows more about tropical fruits, bread, chicken, pasta, and hot pepper sauces than he ever did. He has written about these things. He has written about how he has researched, and taught about, these things. But he's not an isolated individual. He never works alone.

John Eyles is research professor in environmental health at McMaster University. He chairs the Environmental Health Program funded by Environment Canada, through three Canadian research councils—social sciences, natural science and engineering and medicine—and by local sponsors. He is director of McMaster Institute of Environment and Health. He is also Professor of Geography and cross-appointed in the departments of Clinical Epidemiology and Biostatistics, Sociology and the Centre of Health Economics and Policy Analysis. He is author or coauthor of some 150 books, peer-reviewed journal articles, and technical reports in the health and social sciences fields. His research interests include the nature of community decision-making structure; community responses to environmental events; the relationships between environmental quality (and degradation) and human health and between human values, health, and environment; and the development and application of appropriate methodologies in applied settings.

Lawrence Knopp is professor and chair of geography and director of the Center for Community and Regional Research at the University of Minnesota-Duluth, and adjunct professor of geography at the University of Minnesota—Twin Cities.

Janice Monk is executive director of the Southwest Institute for Research on Women (SIROW) and adjunct professor of geography at the University of Arizona. She has coedited or coauthored seven books and published about seventy articles, mostly dealing with gender themes in geography or with issues in geographic education. Among these works are *The Desert Is No Lady: Southwestern Landscapes in Women's Writing and Art* (coedited with Vera Norwood), and *Full Circles: Geographies of Women over the Life Course* (coedited with Cindi Katz).

Pamela Moss is associate professor in the Faculty of Human and Social Development at the University of Victoria, Victoria, B.C., Canada. Her interests in commonplace activities and the mundane has led her to explore the everyday lives of low-income women, older women living with arthritis, and women diagnosed with chronic illness. Her writing considers the themes of body, self, and labor in numerous contexts. She is also an activist in local feminist community politics.

Robin Roth received her B.A. from the University of Victoria in 1997. She then held an internship for the University of British Columbia's Sustainable Development Research Institute (SDRI) in northwest Thailand where she helped a local Ph.D. candidate conduct biodiversity research. Interested in explicitly combining her personal/political passions with her academic ones, she is currently pursuing a Ph.D. at Clark University.

Rachel Saltmarsh researches recent cultural changes in coal-mining communities in South Yorkshire, England. Rachel was born and spent her childhood in Rossington, a coal mining village in South Yorkshire. Her experiences of how these working-class communities were represented from within academia developed an interest in the politics of knowledge construction. Using ideas of situatedness and autobiography, Rachel has previously written about her experiences within her community and academia. Rachel is currently completing her doctoral thesis at the University of Wales, Aberystwyth.

PLACING
AUTOBIOGRAPHY
IN GEOGRAPHY

Writing One's Life

Pamela Moss

> *There was only one letter in the mailbox—and it was for Sophie. The white envelope read: "Sophie Admundson, 3 Clover Close." That was all; it did not say who it was from. There was no stamp on it either.*
>
> *As soon as Sophie had closed the gate behind her she opened the envelope. It contained only a slip of paper no bigger than an envelope. It read: "Who are you?" . . .*
>
> *Perhaps she should see if any more letters had arrived. Sophie hurried to the gate and looked inside the green mailbox. She was startled to find that it contained another white envelope, exactly like the first. But the mailbox had definitely been empty when she took the first envelope? This envelope had her name on it as well. She tore it open and fished out a note the same size as the first one.*
>
> *Where does the world come from? It said.*
>
> —Jostein Gaarder, *Sophie's World*

Recognizing Autobiography in Geography

I came to autobiography as a form of expression in geography after I came down with chronic fatigue syndrome (CFS).[1] I knew of autobiography before; I had studied it and knew a little of the litera-

Thanks to David Butz for his critical, insightful comments, and to an anonymous reviewer. I wrote this chapter while a Visiting Professor in Human Geography at the University of Vienna, Austria. I thank the Department for their support.

1. Chronic fatigue syndrome is a brain disease that has myriad effects on the body, including, but not limited to, debilitating fatigue, cognitive impairment, flu-like symptoms, a long-term low-grade fever, muscle and joint pain, photo- and noise sensitivity, nightmares, memory loss, and balance problems. The disease also

ture. I even wrote "The Autobiography of Pamela Moss" in fourth grade. At the time, I was sorely disappointed in having to report my life as it was. Nothing exciting had happened to me—I had never flown in an airplane, I had never had a pet, I had never moved. I remember when I handed in the paper how anxious I felt because I knew, just any day, something exciting was going to happen, and it wouldn't be recorded in my autobiography. I was right. I broke my arm sledding, on Valentine's Day. I wore a cast for six weeks.

But it wasn't until I was ill that I linked the notion of autobiography, not with recording my life, but with my lived spaces, my own geography. Words are formidably inadequate to explain the impact of CFS on my life. *Everything* changed, from what I ate and how to whom I interacted with and why. Saying I was "tired" couldn't encapsulate my combination of symptoms, but that was all I could think of to say when someone asked me how I felt. In CFS circles, analogies are useful in describing the intensity of the illness, as for example, to call CFS "chronic fatigue" is like calling pneumonia, "chronic coughing." I've never had pneumonia, but I have a new appreciation for it.

My sleep, terribly disrupted, became the center of my existence: I organized my waking hours around resting. Initially, I needed some twenty hours of rest a day, most of the rest of which I spent in a state of lethargy and agony. I'm still recovering but now need only fourteen hours of rest a day, most of which is sleep. My brain fog has lifted and returns only when I overextend myself. At a point during my illness when I could think about something other than support and subsistence, I turned to trying to understand what material conditions and life circumstances led me to my position as an ill, female academic interested in work and home environments of women with chronic illness, that is, who I was and where my world had come from.

Academically, for me, the tone and gist of this venture were not entirely new. I have always tended toward microscale studies in an attempt to place social beings and experience in context while avoiding the offense of methodological individualism (e.g., Moss

---

goes under the name ME *(myalgic encephalomyelitis)* and CFIDS (chronic fatigue immune dysfunction syndrome). Although not permanent, the symptoms last for months, and even years, without complete recovery.

1995a, 1997a, 1997b). I have also been interested in how we position ourselves as researchers in the research process (Moss 1995b,
1995c). In much the same way that I have used the stories of the
women I have talked with over the years,[2] I decided to use myself as
a source of information. It was not so much to compare my life with
theirs; rather, it was more to juxtapose mine with theirs, in a more
collective sense, not point by point, oppression by oppression. I
wanted to use my experiences the way I used theirs—to elaborate
empirical links with concepts, to contribute to critically informed
uses of the individual in political economy studies, and to shed light
on the dearth of feminist analyses of materiality (both economy-
and matter-based) in geography.

As for me, I think I lead a rather ordinary, mundane life.[3] No matter where I am, I do much work, both paid and unpaid, visit with
family and friends, read murder mysteries, and go to movies—lots of
movies. By invoking this rather meek description of my daily life, I
don't mean to dismiss my privilege; on the contrary, I wish to understand such privilege in the context of my own ordinariness,
mundaneness, for it is through this invisibility that the subtleties of
power express themselves—either in being oppressed or in being an
oppressor. For example, when I was first ill, I was dismissed by bio-
medical practitioners and defined as a typical, aging, "hysterical"
woman with relationship stress who needed a job without the demands of thinking so much.[4] At the same time, my class, race, and
occupation facilitated my recovery from CFS precisely because of
the biases toward white professional positions, such as a university
professor, with regard to, for example, the employer carrying both
short- and long-term disability insurance policies, the possibility of
me negotiating a reduced workload because of the type of work I
do, and my ability to articulate in administratively intelligible terms

2. This includes a set of interviews with home support workers in the Greater
Victoria area that were never transcribed or analyzed.

3. Some of my closest friends disagree, Ann in particular. But this is perhaps because we are all located similarly, caught up within the same sets of social processes
that shape our lives.

4. This advice was much the same as women a hundred years ago received when
dealing with similar bodily sensations (see Herndl 1993). Had I been in this situation in 1900, I would probably have been diagnosed with neurasthenia.

the connection between my illness and my unsafe workplace.[5] These experiences were sculpted out of my entanglement within the fluctuating structures of capitalism, racism, and patriarchy. As experiences, they need excavation in light of other mundane and ordinary lives because it is here in the very fibers of the everyday where society reproduces itself, materially and ideologically.

The purpose of this introduction is not to provide an overview, either cursory or exhaustive, of geographic works utilizing autobiography in its various forms, or even autobiographical works elaborating the contours of our lived spaces. Neither is the aim to contextualize the pieces included in the collection, or to give a critical assessment. And I don't want to go through reams of literature to show you that I've read what I was supposed to read. Instead, I set the goal of this introduction as a simple triad: to reintroduce autobiography into geography, to provide empirical examples of three primary ways geographers use autobiography in geography, and to abet discussion in critical circles about constructive ways to use autobiography. The book's very existence meets the first two objectives, and the third can only be assessed at some later time. But this doesn't mean I have nothing to say. As a way to introduce this collection, I first review my role in shaping this book as an academic project, taking stock of the general trends of autobiography generally and autobiography in geography specifically, and then delimit various types of uses of autobiography.

Shaping the Book

I directed the content of each contributor's piece. Although this is not my usual approach to gathering pieces of work from others,[6] I found that when I mentioned autobiography or autobiographical

5. I have been systematically and personally harassed in my workplace. As of July 1999, the university granted my request for transfer, to the Faculty of Human and Social Development.

6. I prefer to solicit manuscripts on a general theme and then make sense of the submissions. I realize that this is *not* the preferred style of edited collections popular in geography right now, one that demands a tight focus and recondite integration of chapters.

writing several people were rather put off.[7] I was a little surprised with such a reaction given what I saw as a routine approach, particularly through the use of anecdotes, in a wide variety of geographical presentations and writings. This directness proved to be a useful, expeditious way of recruiting geographers to contribute manuscripts. I set out initially to find geographers at various stages of their careers in the academy. Robin Roth and Rachel Saltmarsh were both undergraduates at the time they wrote these pieces, and are now both in doctoral programs. Ian Cook was a graduate student when I contacted him and is now in a university teaching position. Kevin Archer, David Butz, Lawrence Knopp, and I are associate professors, somewhere between trying to establish ourselves academically and creating well-established careers.[8] Anne Buttimer, John Eyles, and Janice Monk are full professors who have all had a major impact on their respective research areas as well as on the discipline more generally. Once I found these people, I asked them to write about themselves, their careers, their lives. The specific content of the chapters emerged through discussion, mostly via e-mail, of the different aspects of autobiography.

Still, in the end, the structure of the book and the framework for the content are my conception, my vision, my project. This has both advantages and disadvantages. I benefit because I get to make the point I want to make: autobiography in geography is used historically, methodologically, and analytically. Unfortunately, the range of the types and uses of autobiography are probably not represented.[9] Nonetheless, I think that these papers indicate many of the questions we as geographers are asking ourselves at the turn of this century as well as the issues we variously face during our careers.

Within Autobiography

Autobiography as a critical literary genre reemerged in the early 1980s (Jelenik 1980; Olney 1980, 1988; Jouve 1991). In light of the

---

7. Not everyone was put off. Many people I have come into contact with have been enthusiastic and supportive. I hope they are not disappointed.

8. Knopp was promoted to full professor during the course of this project.

9. Psychoanalysis is a glaring example of the unrepresented.

rising interest in both daily life and the self as well as feminism's claim that the personal is political, autobiography re-surfaced as less than the monolithic category it had been previously. Even after having undergone considerable scrutiny, the boundaries of autobiography in literature dominate critical discussion. This is particularly true for the autobiographies of women even though, paradoxically, it has been feminist writings that have moved the discussion of autobiography forward and contributed to the development and acceptance of the genre once again.

Having once been marginalized as an unacceptable issue for discussion in literary studies, women's autobiography is now receiving warmer receptions and critical attention (Neuman and Kamboureli 1986; Benstock 1988; Brodzki and Schenck 1988; Heilbrun 1988; Personal Narratives Group 1989; Gilmore 1994; Coleman 1997). Historically, women's autobiographies are being used to detail the transformation of women's place in society in the past two centuries (e.g., Goodman 1986; Sanders 1989; Fowler and Fowler 1990). As well, the lives of marginalized women are being made visible through collections of personal writings—giving both contemporary and historical voices to various groups of women (e.g., Allen 1990; Schreiner 1992; Wood 1994; Brewster 1996; Native Women's Writing Circle 1996). Although women's writings are not always explicitly autobiographical, they all provide insight into women's lives. Often women's writing is categorized by nationality, giving a relatively clear axis around which women can identify themselves to women around the world (e.g., James 1990; Abu-Lughod 1993; Domb 1996; Hiroko 1996; Holmes 1996). Sometimes there are spatialized versions of women's autobiographical selves in their writing (e.g., Norwood and Monk 1987; Brown and Goozé 1995).

Although writings in and on autobiography have been located primarily in literary and cultural criticism (e.g., Bell and Yalom 1990; Miller 1991; Wolff 1995; Baisnee 1997), there is increasingly more attention paid by disciplines with strong bonds to geography, principally anthropology and sociology (e.g., Denzin 1989; Jackson 1990; Visweswaran 1994a; Geertz 1995). Feminism especially has fueled the desire to bring together critical writings of personal experience (see Scott 1991; Okely and Callaway 1992; Stanley 1992). Claiming the personal is political set into motion a path for white,

North American women to raise their consciousness about their own (oppressed, exploited, marginalized) positions in society. This rallying cry soon became separate and distinct from the call to theorize different aspects of women's lives, their labor, their sexualities, their writings, their culture, their identities, so that women could benefit collectively from insights from other women, so that they could extrapolate an understanding into areas where women could effect positive political and social change.

Yet underlying the call to go "beyond" the personal is the assumption that there is not theory in personal experience: we can theorize about autobiography (see, e.g., Olney 1988), but autobiography is not theory. As long ago as 1983, however, Stanley and Wise (1983, 11) said that to argue there is a need to go beyond the personal is "a false distinction between structure and process." They suggested using less theory and more experience to understand and explain the process and structure of oppression. A little later, Currie and Kazi (1987, 93), too, tried to resolve the tension between the personal and the political by arguing for the need to develop research that is "not just 'personal' but which relates the 'personal' to the social as a prerequisite for the unity of theory and practice." The "personal" cast in this way has been both motivator and theory for women of color in explaining their experiences of gender and race, and for lesbians in queer theory, their gender, sexuality, and race (e.g., Mohanty, Russo, and Torres 1991; hooks 1992, 1997; Hart and Phelan 1993; Grosz and Probyn 1995).

Within Geography

Autobiography and geography are not strangers. Historically, autobiography has been used to chronicle geography as a discipline. In fact, in the vein of Anne Buttimer and Torsten Hägerstrand (1988) and the sporadic interviews in various geography journals, autobiography is one of the most fruitful ways to access the process of building a discipline. And, to be sure, the history of geographic thought relies heavily on multiple forms of autobiography, including biography (e.g., Smith Forthcoming), mapping career paths, (e.g., James 1981; Barnes 1996) and contextualizing the emergence of intellectual trends (e.g., Harvey 1973; Johnston 1991; Cresswell

1998). Explicit uses of autobiography became more popular after geographers ventured into phenomenological inquiry, at the urging of Anne Buttimer (1976). Such journeys peaked in the 1980s (Browning and Borowiecki 1982; Buttimer 1983b, 1986; Bennett 1984; Billinge, Gregory, and Martin 1984), with some works appearing after a decade of their author's collecting autobiographical information on geographers' careers (Buttimer and Hägerstrand 1988), and then seemingly disappeared (one exception being Porteous 1989). Even with the influx of qualitative methods in geography throughout the late 1980s and early 1990s that showed the extent to which, for example, thick description, depth interviews, and life histories can shed light on the minutiae of daily life (e.g., Eyles 1988; Eyles and Smith 1988), autobiography remained "missing."

Though autobiography as a category may be missing in geography, the use of autobiography is not. Methodologically, autobiography in geography can be closely linked to qualitative methods, particularly ethnography, and to social theory, as it has been in other social science disciplines. Recent incarnations of autobiography in geography have been predominately limited to methodological queries about reflexivity and positionality, particularly in the feminist literature (e.g., England 1994; Gilbert 1994; Katz 1994; Rocheleau 1995), but not exclusively (e.g. Abramson 1993; Keith and Pile 1993). Others have more firmly and explicitly located their inquiry within autobiography (e.g., Moss 1997c, 1999; Valentine 1998; Bondi 1999).

Autobiography is not only a source of data or an approach to research. Analytically, geography has yet to explore the myriad ways autobiography can assist in critique and theory building. In some ways, the questions raised and interrogated in the literature on subject, identity, and subjectivity come closest to the path autobiography can blaze in analysis (e.g., Keith and Pile 1993; Bell and Valentine 1995; Pile and Thrift 1995; Duncan 1996; Yaeger 1996; Kobayashi 1997). The postmodernist claim that the "author is dead" and, its corollary, "there is only text," have been provocative agents in this literature, particularly because such proclamations came at the time subalterns were creating their own voices and attempting to insert them into the production of knowledge. Yet given such slogans and reactions against them, there seems to have been little thought as to how the subject is being reconceptualized and remapped *autobiographically* in geography.

These autobiographical excursions have not been categorized as autobiography in geography's disciplinary history. Instead, autobiography has been subsumed in either phenomenological studies or the documentation of the history of geography (as examples, see brief overviews of the history of geography in Cloke, Philo, and Sadler 1991 and Unwin 1992). As such, autobiography has really only ever been cast as either a bias-screening method or a source of information. This somewhat obvious neglect should not be read as a benign oversight; rather, it is a refusal to accept ourselves in our multiple capacities in the construction of geographical knowledge. We need to accept what we do and scrutinize its meanings. Geographers continue to use autobiography in their data collection methods (life-histories, in-depth interviews in everyday lifeworld studies, full participant observation, the phenomenological method, and ethnography) and in analysis (reflexivity, positioning, situated knowledges, psychoanalysis) without directly mentioning the links to what other disciplines refer to as the autobiographical. Not explicitly recognizing such links glosses over our understanding of the various ways autobiography is (and can be) used in geography.

With the rise of critical reflexivity and social theory in geography, the time has come to rethink these and other uses of autobiography and to think about them more self-critically. Struggling to survive the devastating political effects of identity politics, it seems appropriate to turn our attention toward the construction of "I" rather than toward the perpetuation of the falsely constructed omniscient "eye." Critical geographical analysis is now three decades old and has reached a stage where analysts need to place themselves critically in the research process and the construction of geographical knowledge. Self-scrutiny, individual and collective, can contribute to a better understanding of and provide clearer insight into who we are and where our world has come from.

Delimiting Autobiography

In what follows, I introduce a range of uses of autobiography I've found in my readings—readings that forced me to rethink what it is about autobiography that motivated me to link who I am with the world around me. This classification is only heuristic. I don't intend it to be used as a bounded assortment of the possible uses of autobiography in geography; rather, I wish only to provide a few examples of some of the uses of autobiography while reintroducing autobiography into geography. I hope this assists me in reaching my objective of abetting critical discussion of autobiography in geography.

Confession. Self-absorption. Anecdote. Navel-gazing. Most writers of English-language autobiographies record their stories in an accessible format, as, for example, diaries, journals, memoirs, books. Upon reading private writings, those intended to be read only by the writer, one encounters there an apparent air of confession, of secrecy. After having revealed a self so extensively, the words beg from the reader forgiveness, redemption, absolution. Just as demeaning to autobiography as attributing to it confessional airs are popular conceptions of autobiographical writing: that it is an expression of excessive indulgence coupled with self-centeredness and an inflated notion of the importance of one's self. Drawing on experience for either explanation or justification is often derided as merely anecdotal, with the implicit assumption that such a category of information is less valid than others. To engage in autobiographical writing is perceived as tantamount to wallowing in idleness, to engaging in misanthropic activities, to gazing at one's navel.

Authority. Legitimacy. Truth. Paradoxically, these confessional airs of a written statement—other than a diary or journal—supposedly lend credence to what an author writes. In addition to being rooted in different spheres of knowledge, and subsequently making different types of truth claims (à la Habermas 1981), this process of lending credence is no doubt highly gendered, since historically in North America, women have written the more "confessional" private diaries and men, the "truthful" public works of science, philosophy, and medicine.[10] As well, this process is sexualized in that heterosexual stories are easily told whereas homosexual ones often remain hidden;[11] and racialized in that nonwhite oral traditions and modes of communication are marginalized in mainstream North America while white ones are canonized. A public airing of "secrets" then would be taken as statement of fact, for presumably no one would air her or his *own* "dirty laundry" unless it were true.

This access to truth via experience is a common way for individuals to give more weight, more significance, more meaning to their opinion. All of us have at some time or another been in a situation where someone at the table or in a conference presentation says, "that's not my experience." This phrase, at times innocently wielded, at others, strategically brandished, signifies an attempt to claim authority and seek legitimacy. Such a claim, invoked through culturally sanctioned verbal and textual devices, paves the way for the truth to be extolled. We are then forced into a position of either acceptance through nonconfrontation or argument, with the insinuation that the other person is somehow lying.

This sort of autobiographical technique is often used in academic writing as a way to exude warmth and personality or to give the author a smooth transition, sanctioned justification for introducing a subject or a pleasant parting for the reader to ponder (e.g., Chouinard and Grant 1995; Grosz 1995; Harvey 1996; Domosh 1997). But all autobiographical anecdotes cannot be read, nor should they be, as a natural way to express things—underlying meanings need scrutiny, they are not to be merely dismissed or to remain unchallenged.[12] And even those who refuse the label autobiography still must recognize their link to these techniques (e.g.,

10. This categorization of autobiographical writing supports the arguments that geography as a discipline is masculine (see Rose 1993).

11. For a look at how this process affects lives in filmmaking in Hollywood, see *Celluloid Closet*. This process is imbued with various sets of power that are historically and geographically specific. Other examples of "hidden dirty laundry" include illness, ability, race, and ethnicity.

12. As a complement to these points, see Stanley's (1992, 181–213) discussion on the construction of truth through autobiography.

Shrethsa 1995). For example, Edward W. Soja's (1996) description of the people living across the street from him in Amsterdam, to which he has access through his front window, is a case in point. He constructs lives, authoritatively, without interaction, without permission, invoking his claim to the truth through his method of recovering the "geohistory" of a neighborhood street by reading the landscape. This authority arises out of his claim to know because he lived there for six months. This incident would not be so jarring if Soja hadn't the habit of claiming legitimacy and therefore authority through what Roland Barthes (1977) calls "biographemes," as, for example, when Soja justifies his choice of bell hooks's writing (e.g. 13, 86), recounts meetings with Henri Lefebvre and bell hooks (33, 104), refers to Mike Davis as "Mike" in the text (229), thanks Jansen as "Adrian" in the text for revealing an Amsterdam to which he had no access (292), describes himself in larger-than-life terms (283), and praises melodiously and reproaches sardonically Hayden White (173–74, 182). Once on paper, published and embellished, these claims, no matter how sincere, how inculpable, have a (strategic) legitimacy that lends credence to whatever *else* Soja has to say.

Life story. Self portrayal. This notion of legitimacy is also central to the genre of autobiography. Who better to recollect a life story than the main character of a daily journal, diary, or memoir? Authorization of the truth about one's life is assumed through an "I," not a "she," a "he,' a "we' or a "they." In this sense, conceptualized as someone's life story, autobiography overlaps most with its companion, biography, and is used primarily to chronicle geography as a discipline. Yet, as Liz Stanley (1992) argues, to write biography, one must know her/his own life story. Further, if Stanley is indeed correct, then we also have to accept that in order to write one's own life story, one must know how an "I" fits with other lives, the "shes," "hes," "wes," and "theys." In geography, there has been perhaps even less biography than autobiography recently; biography too, perhaps needs to be resuscitated (recent biographies include Paterson 1984; Fleming 1988; see also Smith forthcoming). A more explicit merging of the two forms would bring new types of work to the fore. For example, as a companion to John L. Paterson's (1984) book on

David Harvey's geography, it might be useful to see David Harvey's autobiographical take on *David Harvey's Geography*.[13]

Geography can probably benefit most from autobiography, in its genre form, through autoethnography. Autoethnography is variously defined as a strategy of colonized peoples to reclaim their own history (Pratt 1992) or an account of one's own life in the field (Paules 1991). And, although mainstream anthropology refuses the title "autobiography" (Clifford 1997), much of conventional anthropological writings are autobiographical for they are ethnographic accounts of being in the field, even if written in the third person (e.g., Malinowski 1922; Mead 1928; Powdermaker 1966; Rabinow 1977).[14]

Writing autobiography as both art and life elicits impassioned responses. In feminist literary criticism, the move to excise fiction from autobiography, or fact from the art of writing (see debate in Stanton 1984 and Brodzki and Schenk 1988), in order to gain more legitimacy, is parallel to the debate over truth and fiction. For example, for Kamala Visweswaran (1994b), the line between the two are permeable and retractable: on the basis of her own experiences in India, yet without ever recounting any one experience, she constructs everyday life stories about her collection of saris. This type of writing in geography is rare (e.g., Moss 1993; Soja 1996, 237–79), but it is one that could be usefully recovered and strengthened.

Lifeworld. Daily life. Experience. Most geographers using qualitative methods, from a variety of theoretical perspectives, including humanism, feminism, marxism, and poststructuralism, are familiar with lifeworld studies. From the philosopher Husserl, through the sociologist Schütz, to the social theorist Habermas, to the geographer Buttimer, lifeworld has come to be known as the experiential life of an individual, the realm where taken-for-grantedness is taken for granted. It is only through critical reflection and appraisal that

13. Such autobiographical reflections on one's own set of works as it contributes to a specific field of study are increasing (for examples in geography, see McGee 1995; and Tuan 1998).

14. See also Sanjek 1990 on the link between recording fieldnotes and the construction of anthropological knowledge.

this world can be consciously discerned. Lifeworlds are accessed through interrogating daily life activities, for example, labor, communication, and meaning (Dyck 1989; Burgess 1990; Katz and Kirby 1991).

Experience has been a little more problematic, both inside and outside geographical writings. Walter Benjamin (e.g., 1978, 1979) spent much of his career trying to understand and explain how experience feeds into theory (for interpretations of his writings, see, e.g., Lunn 1984; Frisby 1985; Nägele 1991; MacCole 1993). He also was interested in how to extract the bonds between personal and professional writing. Feminism in particular has shed new light on how personal experience can be used in geographical research methods as well as how such experience can provide insight into the analysis of social, political, cultural, and economic phenomena (e.g., McDowell 1992a, 1992b; Nast 1994; Nagar 1997). This revived interest in merging academic and personal writing, or the critical and the autobiographical, arises in response to (a) problematizing how a subject is represented in text causing a spate of textual and conceptual justifications for "who the author is" (Miller 1991a, 20),[15] and (b) popularizing the move away from abstract theory toward concrete specificity (Wolff 1995, 45; in geography, see Rose 1993 and Kobayashi et al. 1994). The reinvigorated use of autobiographical writing addresses the question Who speaks *as* whom? which partially displaces the questions Who speaks? and Who can speak for whom? As well, clusters of abstractions attempting to explain everything within one theory have fallen out of favor, replaced by specific theories for specific processes. For North American feminists these two intellectual moments are especially crucial as they struggle to incorporate the insights of poststructuralist thought on difference and diversity into frameworks rooted firmly in personal experiences of power.

This movement toward specificity as applied in autobiographical writing presents a new set of difficulties. Joan W. Scott (1991) cautions against making natural any individual experience (for this

15. She argues that this is not the same as the postmodern crisis of representation. It is a compounding of this crisis by including the multiplicity of possible authors.

tends toward essentialism) and argues for a commitment to categories of analysis that are contextual, contested, and contingent.[16] Also, calling into question the notion of referentiality, whether or not the text corresponds to a "real" point of truth, is imperative, for it is here where one queries legitimacy—even if only surficially. Although blurring the lines between fact and fiction in academic writing is a point of contention, the question of truth correspondence in autobiography isn't. There still is an assumption that when invoking autobiographical information, there has been an embodied experience the author is relating. Although theorists can argue that to question referentiality distorts the picture of the autobiography as a story (Lionnet 1988, 260), when writing one's life, referentiality is exactly what is at stake. The question then becomes Does referentiality have any imagined spaces?

Reflexivity. Positionality. Two principal ways geographers have used autobiography in their approaches to research have been to reflect on where they are located in the web of power relations constituting society and to utilize this positioning as a mediating relation in the interpretation of information gathered through the research process. These arguments have arisen primarily from engagement with feminist critical cultural theory, as, for example, through Gayatri Spivak's work. What is interesting, however, is that Spivak does not effectively integrate her positionings into her analysis. For example, setting her own experiences into a context of Indian nationalist history, Spivak (1989) shows the process through which some histories of the "Other" are appropriated by neocolonialists. By using a few of her own reflections in a critical reading of a text, she is able to show the important role mediation of historical events plays in shaping how historical texts are written. Rather than simply reading the texts with their masks (in her example, the masks of nationalism, internationalism, secularism, and culturalism) as documented history, she argues that remaking a history that is truly postcolonial would involve a "persistent critique, unglamorously chipping away at the binary oppositions and continuities that emerge continuously in the supposed account of the real" (Spivak

16. Arguments developed more in relation to space can be found in Mouffe 1995 and Jones and Moss 1995.

1989, 288). These masks indeed are the manifestations of Spivak's embeddedness in her own autobiography. Yet she stops short of an impassioned plea for a critical reflection of our own political positionings. An autobiographical critique, it seems to me, must challenge the process of writing history, not just its end product. In not fully integrating her own political positioning into the analysis, Spivak merely draws on her experiences to legitimize her approach to how colonized histories are and can be written through politically positioning "Other" as subjects in historical texts.

In geography, most of the successful attempts at integrating positionality into analysis have been in feminist analyses (but see also Bonnett 1994, 1996; and Routledge 1996). Cindi Katz (1992) effectively moves back and forth between her own experiences as a researcher and the theory informing her analysis. And, the biographemes she summons forth highlight the theoretical points she discusses rather than attempt slyly to enhance the credibility of what she has to say (Katz 1992, e.g., 503, 504, 505, and especially 507 with reference to *"I'm* the other Cindi!"). There are many points where geographers could more readily entangle their positionality in their analysis: Julia Cream (1995) in her research on the intelligibility of the pill; Laura Pulido (1997), in her participant observation work with environmental activist groups in South Central and East Los Angeles and Ann M. Oberhauser (1995), in her work on household strategies in rural Appalachia. This is not to denigrate the work that is done; rather, I only identify these works as examples where autobiography could add substantively to the interpretation of the information gathered.

Criticism. Analysis. Theory building. Analytically, autobiography is not developed in geography. Yet it has the potential, as we are seeing with critical turns in analysis in other disciplines, especially literary criticism, cultural studies, sociology, and anthropology, to be a powerful critical tool for analysis. Disciplines intersecting with geography, especially those that entertain the notion that the self needs to be recast in the face of the postmodern critique that the subject is dead, have turned to a critical analysis of how to use autobiography (Derrida 1985; Eakin 1985; Elbaz 1987; Lejeune 1988; Ashley, Gilmore and Peters 1994). With the reclamation of women's writing in literary criticism (e.g., Greene and Kahn 1993), autobiography has taken on roles other than popular literature, information source, methodological approach, and writing genre; it is now part of criticism itself.

Mary Ann Caws (1990) uses what she calls *personal criticism* (developed further in Miller 1991):

> a willing, knowledgeable, outspoken involvement on the part of the critic with the subject matter, and an invitation extended to the potential reader to participate in the interweaving and construction of the ongoing conversation . . . even as it remains a text. The experience is open and fluid, as is the transcription of the implicit conversation. . . . [Personal criticism] is composed of an unshakeable belief in involvement and in coherence, in warmth and in relation. (Caws 1990, 2–3)

She engages in personal criticism to critique biographies of the "Bloomsbury" women—Vanessa Bell, Dora Carrington, and Virginia Woolf—their self-representations, and the choices they made around their loves, art, and sexualities. Although not thoroughly autobiographical, Caws does incorporate who she is in the critique by stating opinions openly about decisions these women made and playing them off her own interpretation of their biographies, self-representations, and choices.

This type of inclusion of our own personal lives should be not only substantive, but also analytic. Janet Wolff (1995) exemplifies what Caws and Miller have outlined as personal criticism. She deftly unravels the problematic notion of referentiality in an essay on "Death and the Maiden." After having watched Woody Allen's *Crimes and Misdemeanors,* she is troubled by the use of "Death and the Maiden" just before the female character returns home to her death, arranged by her lover. On her second viewing of the film, she finds out that the music, although Schubert, was not "Death and the Maiden." This shielding of a murder in an exchange of music focuses the viewer's attention away from death toward the intensification of the emotions involved in *deciding* to kill. From this autobiographical account, however, she is able to argue that the status quo is being served by patriarchal reinventions of classic cultural entities that perpetuate images of (and justifications for) violence against women. Her recovery of the "truth" of the corresponding reality is a political act, a integral part of her own autobiography, in that she insists that we cannot forget that the "Maiden" indeed dies, each and every time, no matter how the text of the lyrics are presented, visually or aurally, to protect us and redeem them from violence.

Personal criticism is not the only way autobiography can be used analytically. Autobiography has proven useful in other disciplines to define new directions, to provide theoretical insight, and to promote a continual destabilization of the formation of theory. In "Sexing Elvis," Sue Wise (1984) shows how the disjuncture of her own experience of Elvis Presley differed radically from media representations constructed primarily by men for men. She chastens herself and other feminists for taking these representations as "objective fact" without ever asking how women experienced either Elvis or rock music. She uses her own experience to point cultural criticism away from the masculinized norm into a direction that raises different questions about popular cultural icons that bring into focus women's everyday understandings of the world, their lifeworlds.

In "Blondes," Eva Podlesney (1991) says that she bleached her hair blonde "not only in a counter-cultural *homm(e)age* to the Hollywood studio stars, but also, glibly, as an anti-culture critique of the manipulation of women's images in the name of male desire" (70; emphasis in original).[17] In constructing her argument, she uses her own rumination of Madonna's hair color(s) to revisit the construction of the Hollywood blonde phenomenon and women's desire to *be* the blonde image. After reviewing the construction of the careers of such blonde "bombshells" as Jayne Mansfield, Dorothy Malone, Marilyn Monroe, Kim Novak, and Shelly Winters, she argues that Madonna appropriates this soft, lustful, wispy, seductive image and replaces it with a muscular, rigid, icy, aloof, blonde image.[18] Wearing tightly wound bondage costume and a long, blonde ponytail pulled

17. I do not want to diminish Podlesney's arguments, nor do I want to offer a blatant misreading of her work. She uses the act of bleaching her hair blonde as an introduction to the essay on "Blondes." Although Podlesney does not argue for any use of the autobiographical, methodologically or analytically, it still does not detract from the usefulness of this article in showing *how* autobiography can work analytically.

18. Madonna used the soft, seductive blonde image to presell tickets for her *Blonde Ambition World Tour.*

starkly away from her face in the opening number of her *Blonde Ambition World Tour,* Madonna manipulates the blonde image and subsequently frees *all* women from their relationship to this phenomenon. Although I question the universality of Podlesney's claim of releasing *all* women from desiring to be blonde, she does rely on a collective experience of women in relation to "being blonde," somewhat as Wise relies on the masculine constructions of Elvis as a phenomenon. Yet, rather than identifying negatively with the hegemonic construction, Podlesney uses her experience and her own positive positioning in the blonde phenomenon as theoretical insight. In bleaching her hair, Podlesney identified her own resistance to the blonde phenomenon, and through this theoretical insight, willed her roots to show.

In "A Gender Diary," Ann Snitow (1990) takes us on parallel journeys through how feminism seems to rework itself over and over again along the same divide and how, in her own life as a feminist activist scholar, she seems always to fall on the same side of the divide. She reflects on her experiences in conjunction, not with other women, but with a recurring tension within feminist theory between minimizers and maximizers of difference between women and men. She works through this tension in several contexts. She uses her own experiences to show how this characterization of feminism is not essentializing; rather, it is a way to suggest to feminists how to place their own work on either side of the divide and to assess "how powerful that political decision is as a tool for undermining the dense, deeply embedded oppression of women" (Snitow 1990, 29). She continues to define herself as one who wants to minimize difference between women and men and aligns herself with feminists who have the same preferences, while still being sensitive to each woman's personal material history. She realizes and suggests that feminists should work with the assumption that one's allies can come to the same conclusion from different starting points.

Process. In pushing the analytic borders of autobiography, geographers might find it useful to think of autobiography, not as a record of a life or some aspect of life, but as a life itself. What is autobiography if not process? It is a process, not only of recording, in the sense of documenting, orienting and analyzing, but also of becoming, in the sense of lives, subjectivities, and identities. Though critical self-reflection is not complete, nor is it the only way of knowing, it can be a helpful and workable approach in gaining insight into one's life as well as into the contexts within which one exists. Janice Monk (1997) innovatively, and autobiographically, records her thoughts on writing an introductory piece to a section of a book, *Thresholds in Feminist Geography,* on "representation." She sets out the process she went through in deciding what to write about and why to write it. She discloses, via e-mail messages, circulars, dictionary definitions, photos, script excerpts, poems, and prose, her thoughts on representation, her place in feminist geography, herself as a woman. Monk is able to position (and reposition) herself, analyze the concept of "representation," document a part of the history of feminist geography, and indicate theoretical places where feminist geographers could go, all in one short piece.

Process also involves a sense of connection; connection to those close to us. Edith Sizoo's (1997) project involved collecting sets of women's autobiographies and then pulling out and commenting on common themes. Fifteen women wrote their life stories in conjunction with three other women from different generations. Most wrote about female relatives, but some wrote about close friends and extended family members. As I read through the pieces, I came to think that I actually knew each of the women. I know their names and a little bit about their lives. Each writer wove a tapestry replete with fused life lines from generation to generation. The abundance of particularities, the intricacy of patterns, the integration of lives is astounding. I continually wondered what would happen next, and next, and next. I grieved at the end of each section. I felt incomplete. I thought a lot about the book in the next couple of weeks, trying to figure out why I was so unsettled after having read the book. Then it struck me: it bothered me not to know about the other women in these women's lives—their neighbors, their friends. I began thinking about the other connections in the women's lives. Who else was important to these women? How did each of these women figure into their lives? *Who* are they? Where are their autobiographies? How do they fit in the world? How does the world fit around them?

Writing one's life. For me, autobiography as process implies writing one's life: accessing and documenting the construction of "I" and "me" in context, multiple contexts—cultural, economic, environmental, historical, political, social, spatial—with insight into who "she," "he," "we," and "they" are; positioning, repositioning, and repositioning once again in light of my environments; placing, displacing, and replacing myself in the world over and over and over; designing, maneuvering, reacting, and recording my geography, as I live, through my broken arms and illness, through my (small p) politics, through my writing.

In a parallel fashion, for geography, writing one's life might in many ways be able to augment the continued construction of geography as a discipline, especially with regard to the people who build it: scholars, researchers, teachers, students, practitioners; to tell us who we are in context of our multiple environments; and to give some clues as to where our world comes from. As to the forms writing one's life take, these are as yet untold.

2 Home—Reach—Journey

Anne Buttimer

The opportunity to return to my native land after more than half a lifetime abroad was indeed a welcome surprise in 1991. To return as a geographer, rather than just as a visitor, was something new. For nigh on thirty years there were annual visits, all of which nourished an image of home that highlighted not only the sunnier aspects of childhood but also the contrasts to those of everyday life in my "working" environment at the time, whether Belgium, France, Scotland, the United States, Sweden, or Canada. Challenges along the way have varied widely and traveling companions have been diverse; the common denominator throughout being my geography. Back home as permanent resident, in suburban Dublin rather than in rural Cork, childhood images are now reshaping themselves even as Ireland's landscapes and lifeways are being busily transformed. My life journey so far has been filled with surprises—often pleasant— and few without deep emotional charge.[1]

Childhood in Ireland

Growing up in rural Ireland during the 1950s was, in retrospect, quite a privilege. It was an oral culture, great store set on recitations and songs, solving practical problems "in one's head," and long conversations around the dinner table. Grown-up talk was of land recla-

1. The following paragraphs, and indeed several other pieces of this essay, overlap substantially with those included in a previously published essay (Buttimer 1987a).

mation schemes, rotary milking parlors, farmers' organizations, and Big Bands; good stories were told, tourists began to appear, and there was general admiration for the Foreign Missions. School was not always fun, and during high-school years I was made to feel keenly aware of being two years younger than others in my class. Still I remember enjoying Molière and La Fontaine, Horace and Virgil, Yeats and the Shelleys. I did well in maths and piano, the school play and dance. I preferred poetry to prose and devoured anything I could find on classical history and myth.

Even more I enjoyed getting home from there, participating in debates and social events organized by *Macra na Feirme* (an Irish-style 4-H Club in which my father played an active role). That was a more "real" world, being a farmer's daughter was no longer something to feel apologetic about, as it sometimes was among the posh city girls and the worlds of couth and culture into which my boarding school teachers wished to socialize me. I owe a lot to the Loreto Sisters in Fermoy; study habits and intellectual interests acquired during those years have been invaluable.

Maths and Latin were my favorite subjects in college, geography a fun sideline at first. Girls were not entirely welcome in these courses, but they enabled me to feel at home with authors and ideas far-flung in time and space. Not until I had completed an M.A. degree and taught in a secondary school for a year was my mind made up about a career: it was to become a religious. At first I dreamed of working with Mother Teresa in Calcutta, but later decided to join my older sister on America's West Coast.

Seattle-Tacoma in the Early 1960s

"Not another one from Ireland!" a Seattle University professor remarked one day. Suddenly I felt an exception—in tastes, accent, ideas—in this environment. In general, however, the atmosphere was welcoming and supportive. I discovered psychology and other -ologies, in education courses required for bone fide certification as a teacher in the State of Washington. Within a year came the challenge to pursue doctoral studies in geography as qualification to contribute to Seattle University's "Sister Formation" program—an integrated curriculum for the training of teachers. My special responsibility was to be "social geography": the integrating capstone on a series of courses in the social sciences. It was a thrill to undertake this challenge.

Seventeen years I spent in the Dominican Order, years of emotional intensity, energy, and challenge. The spirit cultivated for over 750 years in this order, *contemplata aliis tradere,* (sharing with others the fruits of one's own contemplation), continues to inspire all of my professional work, long after my formal dispensation from vows. It has also undoubtedly motivated my efforts to encourage international and interdisciplinary dialogue about issues of knowledge and life experience. And all along the way—in Belgium, Scotland, Massachusetts, Sweden, and Canada—there have been Dominican communities where I found welcome and inspiration.

Graduate years at the University of Washington in Seattle during the early sixties opened up fresh horizons, particularly in the social sciences. There was a quantitative revolution underway and regional science was regarded as geography's way for the future. As for what social geography could be, Edward Ullman paternally advised, "We got all that stuff out of our systems long ago—with Ellen Churchill Semple." Much as I admired Ullman, I admired Semple even more, her Mediterranean work having built bridges for me between classical history and physical geography (Semple 1931). Teilhard de Chardin was one of my other heroes at the time, and so I set about discovering other voices on what a social geography might be. With midnight oil I read the work of social geographers in Germany, Netherlands, Britain, and Sweden, and eventually, with support from my advisor, Morgan Thomas, and with generous guidance from colleagues such as Marvin Mikesell and David Lowenthal, concentrated on *la géographie humaine.*

What was most appealing in this literature, mostly untranslated at the time, was its graphic descriptions of regional landscapes, people, and places. The central notion of *genres de vie* (lit. lifeways, or patterns of living) helped to explicate everyday patterns of life through a comprehensive analytical approach involving scrutiny of (a) traditional beliefs, images, and habits, (b) socially taken-for-granted rules in the functional organization of activities, space, and time, and (c) the bio-ecological base of physical milieu, resources, and geographical location as fundamental stage for the drama in different parts of the globe. *La géographie humaine* pointed toward a credible horizon for a human geography that could provide the integrative function I was supposed to perform at Seattle University (Buttimer 1971). In retrospect, too, it resonated to the lived realities of agrarian Europe, my own background. Quite a contrast it was to this new geography developing at the University of Washington, wherein humans were regarded as cost-minimizing, profit-maximizing, Promethean individuals, and the geographer's task was to write the score for a rational organization of space to stage that drama. "Watch out," Marvin Mikesell once warned me, "they will call you Sister Social Geography"(Buttimer 1968).

At the AAG meetings in Columbus, Ohio (1965), I dared to make the case for a French-style social geography, making a special point about objective and subjective social space as a way of elucidating the contrast (and possible conflict) of "insider" and "outsider" perspectives on space and place. That was the meeting where a famous panel on Environmental Perception and Behavior staged its debut (Lowenthal 1967), and it was indeed gratifying to notice that interest in cultural differences in ways of understanding nature and the physical environment was now gaining ground in geography.

Louvain, 1965–1966

"Any interest in Belgium, Sister?" Edward Ullman asked one day on the steps of Smith Hall, University of Washington. A postdoctoral fellowship for the academic year 1965–66 by the Belgian-American Foundation opened up a marvelous opportunity to study philosophy at the University of Louvain and to complete my research on the French classical tradition (Buttimer 1971). Seattle University added a grant of $1,500 so I could glean anything relevant for the setting up of a Geography Department there on my return. Enrolled in the *Institut des pays en voie de développement,* I found a diverse company of Africans, Asians and South Americans, maoists, marxists, existentialists and structuralists, repatriated missionaries, and 1960-type revolutionaries who on occasion would accost me about American imperialism. It was a time for dialogue among visiting students, if not, alas, among Flemish and Walloon within Belgium. Existentialism and phenomenology, hermeneutics and structuralism were in the air; it felt as though many previous "certainties" had all had a sauna. Among visits to France that year, one memorable treasure was the welcome afforded by Paul-Henry Chombart de Lauwe, who opened up analytical horizons on "social space." Aware of the emerging trends in proxemics, mental maps, and perception in America, I felt quite confident about the fresh insights that this concept could afford for empirical research (Buttimer 1969).

Glasgow, 1968–1970

Soon after my return to Seattle in 1966 the prospect of developing a geography department at Seattle University was to fade, as was the "Sister Formation" program for which I had become qualified. The community self-study and renewal movement provided plenty of challenge; each of us was encouraged to follow our own vocations and I was convinced that mine was in the academic world. This conviction led first to Glasgow where for two years, 1968–70, I joined an interdisciplinary team that was critically evaluating planning standards then in vogue in England and Wales (Forbes 1973). Here came the opportunity to test out empirically some of these ideas that had emerged during my *séjour* in France. I studied the "social spaces" of relocated working-class families from slum clearance districts to municipally built housing estates. The aim was to see whether the presence or absence of official planning standards made much difference for the residents, or whether the distance between previous home environment and the new was significant. I documented patterns of territorial identification, activity networks, and environmental images, all of which I assumed would vary not only among individuals, but also between groups, at least between the two most salient social reference groups of working-class Glasgow, Celtics and Rangers (Buttimer 1972).

There was still much to finish on the social space project when the invitation to Clark came in 1969–70. Bob Kates had given me a hint previously but I gave it little thought. Nor had I anything else in mind except a return to Seattle. Meanwhile I had grown quite fond of Glasgow, and lifelong friendships had developed there. Eventually I accepted the offer of a postdoctoral year at Clark, all the time anticipating a return to Seattle.

Worcester, Massachusetts, 1970–1981

Colleagues and students at Clark offered a warm and enthusiastic welcome. There could scarcely have been a better environment for precisely the kinds of academic interest that seemed vitally important then. But I was pining for Glasgow, blaming Worcester for being neither Glasgow nor Seattle, and like my relocated Glasgow "wifees," was finding it difficult to adapt to a new social space. An eager bunch of graduate students quickly joined me for a seminar on social space. They played with the Glasgow data in countless ways, discussed territoriality, images, and activity networks, quite as keen to develop new methods perhaps as they were in learning about either Glasgow or planning standards. A new social space was a-forming, one in which roots were planted, harder to uproot than most others, a decade later.

During 1970–71 I became a student again, in courses with Kenneth Craik, David Stea, and Robert Beck, keen to learn more about environmental perception and to discuss methodological problems. With Dan Amaral and Ben Wisner, the pioneer spirits of *Antipode,* one could air philosophical questions of existentialism and phenomenology; with Roger Kasperson and Bob Kates, issues of values and advocacy planning; the Place Perception Project led by Jim Blaut and David Stea was intriguing; Roger Hart, Gary Moore, and David Seamon organized luncheon meetings where graduate students from both geography and psychology could share insights from their ongoing projects. Then there was a periodic "Faculty Seminar" with a most impressive assembly of psychologists and geographers to which I was also invited. My first impression was one of culture shock: the style of discourse seemed so abrasive. When it was my turn to present the Glasgow study I left the meeting in shreds. Others found the meeting "stimulating!"

The seminar provided an ideal forum to air several of those queries that had been bothering me since the Columbus (1965) session. "Perception" obviously meant different things for different people, but the Clark group tended to define it in terms of cognition. Now this cerebral orientation struck me as being too narrow, especially for the kinds of curiosities that were raised in my Glasgow study. There were emotional, moral, aesthetic, and habitual aspects in taken-for-granted images of home that could not be adequately described in the language of "mental maps." I argued for a focus on experience rather than on perception. Then there were issues of methodological individualism and reductionism; theories borrowed from psychology were for the most part theories concerning individual personality, cognitive development, or motivation. It was not at all obvious that they would yield good insight on the human experience of space, place, and making a home on the earth. What seemed missing was cognizance of social norms and structures, historically sedimented habits and cultural prejudices, power relations of authorities (at various scales), which exercise such discretion over people's attitudes and behavior in space. There was also a certain North American bias in the interpretations of behavior and a tendency to conduct piecemeal empirical case studies with pragmatist leanings in the way conclusions were drawn. In raising such issues with this group I became more aware of the limitations of my own framework, the ways in which I may have imposed my own values on the Glasgow study, and how much I still had to learn.

The early 1970s at Clark were both emotionally stormy and intellectually challenging. After two years abroad, it was a different America I found, quite a contrast in many ways to the one I had known in Seattle during the sixties. Vietnam veterans and antiwar missionaries, environmental enthusiasts and folk inspired by Bunge's *Detroit Expedition;* the walls between academia and society were now more permeable. Internally, too, there were ideological tensions around the birth of new fields of research expertise such as "psychogeography." But *Antipode* was launched and fascinating proxemic patterns emerged in the newly refurbished Graduate School of Geography. What I valued most was the openness and critical attitudes of colleagues and students: exhilarating it was, once the style and lingo were understood.

The Clark years were filled with emotional highs and lows. In retrospect, I now realize the debt of gratitude I owe to friends and colleagues from those years. Despite my nostalgia for Seattle and Glasgow, these years allowed me breathing space for the two enduring intellectual journeys that geography opened up for me: (1) *genres de vie* (ways of living) and (2) intellectual history and philosophy. The first has sought empirical exposure, practical problem-solving and interaction with "real" people in concrete situations; the second was pursuable more or less at home, and it demanded reflection, reading, and theoretical imagination. These two have been like yin and yang for my thoughts and practice, and during the Clark years both found some playspace. Beginning in the fall of 1971 I had responsibility for courses in urban and social geography, as well as in the history of geographic thought, and for the next several years my main research interests were those I shared with my graduate students (Buttimer 1987c, 1992b, 1998a; Buttimer and Seamon 1980). A dedicated lot they were, and we had full support from the school's director, Saul Cohen. With Graham Rowles there were issues of the elderly; with Bobby Wilson and Gerry Hyland, the experience of migrants; with Henry Aay, philosophical questions surrounding the history of geographic thought; with David Seamon, phenomenology. There were questions of environmentally related stress with Michael Godkin, anarchist communes in Civil War Spain with Myrna Breitbart, a quest for geography's new paradigm with Courtice Rose, new urban-social geography with Ruth Fincher, and eventually the lifeworlds of Northwest Indians and those of Bureau of Indian Affairs men with Paul Kariya. By the mid 1970s it was clear that "social geography" had now become an acceptable term in North America.

In 1972, I was asked to write a paper on values by the Commission on College Geography. Having completed what I regarded as a decent preliminary draft, I shared it with some students and colleagues at Clark. "Schizophrenic," one student rapped, "here you claim that existentialism and phenomenology invite us all to become aware of our own taken-for-granted values, and then produce this heap of words without acknowledging one bit of awareness about your own values." Once recovered from the shock, I started all over in a St. Exupery vein, raising questions—sometimes rhetorical—about values implicit in the practice of geography, beginning with myself. I began keeping a personal journal which forced me to reflect critically on issues in my own life journey. Through many tough decisions and emotional upheavals between 1973 and 1979 this journal became a trusty companion. The values paper (Buttimer 1974; see also Buttimer 1996a) became an emancipatory turning point; many contradictions and inconsistencies which I had unmasked in my own life and thought became challenges to confront. Integrity demanded, among other things, transcending those comfortable "isms," "ologies" and a priori biases, seeking the spirit rather than the letter, the ethos rather than the structure in whatever life situation.

The paper itself probed the disciplinary heritage of models, theories, and practices, noting the affinity of certain classical schools of thought with imperialism and earth conquest, the models of mankind that were embedded in conventional theory, and the sometimes absurd transposition of models from one context to another. It also poked at certain contradictions in the sociology of the profession from the socialization of the young, through the rank-and-tenure criteria of academic institutions, to the dynamics of power elites in publication and grantsmanship. Finally, it raised a question about values implied in scenarios about the future—a favored sport in the 1970s—highlighting, of course, the tensions between existential meaning and planned order in the organization of space. The final draft was actually completed in Lund during spring 1973, and the dialogue begun then with Torsten Hägerstrand and his colleagues was to open a whole new phase.

Lund, Sweden, 1976–1988

Genres de vie was immediately recognized as the potential common denominator in my conversations with Torsten Hägerstrand, renowned leader in quantitative and applied geography, when I first met him in 1971. He expressed interest in the Glasgow project and shared some of his own emerging ideas on time geography (Hägerstrand 1970). Recognizing how valuable it would be to integrate a temporal dimension to my analytical approaches to social space, I eagerly accepted an invitation to Lund for spring 1973; my salary allowed Ulf Erlandson to replace me at Clark.

There were discussions on temporality and *genres de vie* with Torsten and his group, but far more fascinating for him personally was my *Values* paper, which he considered to be a "revolutionary document" and promptly set about translating it as well as writing his well-considered response. Far from resenting those rhetorical remarks about Sweden's benign technocracy, he was eager to hear more about the "humanist" tradition, and plans were laid for my return as Fulbright visitor in the spring of 1976.[2]

In preparation for this project I immersed myself in the literature on temporality—Minkowski, Eliade, Bachelard, Merleau-Ponty, Colquhoun and others. The Lund model, it seemed to me, was touching on only one level of spatiotemporal experience, namely, the functional level; its metric was clock-calendar time, and its aim was evidently to elucidate mainly patterns of overt behavior and ways in which institutional forces might actually be caging individuals into paths and trajectories. For comprehensive understanding of environmental experience I felt that one should not only examine images and perceptions of time, the ebb and flow of emotional time, memory, beliefs, and myths, but also the temporal cycles of biological rhythms—the cyclical flows of neurophysiological and ecological processes that envelop daily existence. All the time I was hoping that together we might be able to design conceptual frameworks for the analysis of *genres de vie*, and thereby help to elucidate problems of stress in everyday life.

Once at Lund in spring 1976, however, empirical and methodological questions were to take second place to the more philosophical and practical questions that were then stirring in Academy and Research Councils in Sweden. A seminar on "Knowledge and Experience: Nature Space and Time" was announced and sixty to eighty people, professors and doctoral students from fifteen different disciplines, participated. An ideal setting it was to share reflections about values, the sociology of disciplinary practice, the dream and reality of applied science; people eagerly moved back and forth across disciplinary boundaries.[3]

---

2. Still eager to learn more about temporality, however, I suggested the next time one should arrange not simply an exchange of personnel between Clark and Lund, but rather that both could be together and work out a joint empirical project to be conducted partially in Worcester and partially at Lund. It was possible to arrange for a shared salary that allowed Solveig Mårtensson to spend the first semester at Clark, where we could initiate an empirical project together, with the intention of completing the analysis while at Lund during the second semester.

3. Complete texts of my lectures were precirculated and thoroughly discussed with Torsten's research students, who became group discussion leaders on the occasion of the seminars.

Nineteen seventy-six stands out as a kind of midsummer year in my life. Phenomenology had somehow come to harvest with the "lifeworld" article (Buttimer 1976; see also Buttimer 1994b), my students' dissertations were well under way toward completion, and new challenges were on the horizon with hermeneutics and structuralism. I was graciously invited to lecture in twenty different European departments that year and found great joy in establishing contacts among researchers of different language and cultural backgrounds. My mother died in 1976, having blessed my decision to relinquish my vows to the Dominican Order. It seemed as though from the ashes a new vocation was being born, namely, to building an international community of scholars, where self and mutual understanding could be springboards for better communication between science and society, between scholars and the lived worlds of humanity (see Buttimer 1982). And in 1977 I accepted an invitation for a research position at Lund University on issues of "knowledge integration" and to complete work with Torsten Hägerstrand on the promising volume based on the Fulbright Seminar 1976.

Toward International Dialogue

It was certainly on the strength of the Lund 1976 seminar, and the generous support and encouragement of Torsten Hägerstrand, that I proposed to the Leningrad Congress of the IGU Commission on the History of Geographic Thought to invite autobiographical reflections from senior and retired colleagues as (a) catalysts for dialogue on values in the thought and practice of the discipline; and (b) potential data for an oral history of the field to complement the archival record. For Torsten it also seemed that this approach could yield insight into the general problem of "knowledge integration." Why not try the same experientially grounded approach to dialogue that had worked so well in our 1976 seminar? In 1978 we launched an ambitious international dialogue project (Buttimer and Hägerstrand 1980), one that absorbed the bulk of my energies during the decade 1978–88. The decade afforded many a lesson about symbolic interaction, about language and power, about vested interests in maintaining barriers to communication and mutual understanding between human beings, and about theoretically based objections and even hostility to the very idea of dialogue.

In June 1978 we arranged a seminar/workshop at Sigtuna (Sweden) and invited a number of senior and retired scholars and professionals to share insights from their own career experiences on the subject of creativity and environment. With David Seamon's help a working journal was designed for each participant, so that during the meeting they could record significant events, places, people and projects, and also reflect on experiences that facilitated or hindered their own work. As they left the meeting they were given a short set of specific questions on creativity, place, and horizon. From responses to these questions, as well as from further readings on the subject, I summarized some ideas on phases in the creative process, each phase of which was apparently associated with certain conditions of (a) context (milieu) and (b) communication (Buttimer 1983a).

The 1977–79 period was otherwise one of mixed emotions. The dialogue idea was certainly more welcome in other disciplines and fields than it was among other geographers at Lund. For me personally the most important event of these years was meeting Bertram Broberg, Professor and former Rector of the Lund Institute of Technology, who has shared my life and many of my interests since 1979. Friends, colleagues, and students all appreciate his gifts, not least among which are his evocative sketches and colorful "overheads." On leave from Lund, he was a visiting professor at Brown University, Providence, Rhode Island, 1979–81, while I resumed work at Clark. In 1980, we both received invitations to China and Japan.[4] In Beijing Bertram lectured to geophysicists and engineers and I to the social scientists and urban planners. This was followed by a tour of Japan, where Bertram has been a visiting professor on numerous occasions and where the 1980 International Geographical Congress was held. Our commission symposium was held in Kyoto, a place that has left enduring impressions.

4. It was in summer 1980 also that Gunnar Olsson and Peter Gould organized a symposium at Bellagio, Italy, where geographers of quite different orientations could "dialogue" (Gould and Olsson 1982).

In 1981 Bertram decided to resume his position in Lund and the next eight years were spent there. By the mid-1980s over three-hundred people from thirty-five different countries had contributed to the dialogue process (Buttimer 1986). Colleagues in medicine, business administration, law, architecture, sociology, and literature had found these recordings useful as catalysts for dialogue within their own teaching and research settings. There remained the mammoth task of harvesting insight from this dialogue process about the original challenge of "knowledge integration." From careful scrutiny of the texts—autobiographical reflections as well as published works—insight could be derived on potential common denominators among these scholars—potentially common life experiences that might provide bases for mutual understanding. A trilogy of interpretative themes emerged: meaning-metaphor-milieu (Buttimer 1983b, 1984c, 1993). *Meaning* refers to vocational choice, work, or professional activity; *Metaphor* refers to cognitive style, and *Milieu* refers to both the environmental features of an author's own childhood and formative years and the public interests served by one's research activities. On all three grounds—meaning, metaphor and milieu—there is scope for dialogue and mutual understanding. For Janus-like, the themes face both ways: to the "subjective" experiences of individuals on the one hand and to the "objective" circumstances of power relationships within the changing contexts of their work on the other. They thus serve as potential bases for "zones of common reach" among researchers of widely different disciplinary backgrounds (Schütz 1973).

This trilogy of themes has proven quite useful in contexts beyond history of geographic thought. It served as conceptual frame for studies of environmental perception, migration and identity, and water symbolism (Buttimer 1984a, 1984b, 1985, 1994a, 1994c) and indeed it was exciting to engage in some empirical projects again. Meaning-metaphor-milieu has also provided an organizational frame for courses on theory of science that I offered at Lund during the 1980s and for shorter-term workshops in Poland, France, and Spain and at Clark during those years. To understand environmental experiences of people in cultures other than one's own, I'm convinced, one needs more than a study of environmental perception, or even of social space. One has to learn the language of symbols and metaphor, myth and artifact, around which a sense of group identity is created and maintained. When people migrate from one setting to another, it is often those latent, taken-for-granted aspects of meaning, metaphor, and milieu that assume enormous significance.

For all its heuristic value, the trilogy also became almost like a prison—a template into which I endeavored to "fit" a great variety of puzzles and seek integrated perspectives on each. Then in 1984 my Spanish colleagues invited me to prepare a paper on the history of humanism in geography. This became another eye-opening challenge. Why the breakthrough of certain ideas at particular times, the occasional—gratuitous—outpouring of creativity at certain moments and then the dull plateaus of routine-operational procedures at others? Another trilogy—a mythopoeic interpretation of the Western story—came to occupy my thoughts: Phoenix-Faust-Narcissus. Liberating strains or another prison? Humanism, it seemed to me, could be regarded as the *cri de coeur* (liberation song) of humanity, expressed at times and places where some dimension of life or thought had been ignored, suppressed, or forgotten. "So humanism is something which is crisis induced?" Torsten Hägerstrand teased. Was there not also the gratuitous side, he suggested, the outpouring of creative talent, also appearing spontaneously in history? Just at a time when the whole subject of "knowledge integration" was becoming quite an oppressive one, this new perspective seemed to beckon wider horizons of understanding, or at least some fascinating diversion.

It was more than a diversion. It became for me a liberation. Students responded eagerly to these ideas during a term as visiting professor at Université de Paris I (1986), and during a subsequent one at the University of Texas at Austin (1987). Thanks to numerous exchanges with colleagues around the world, at last it seemed that a "solution" to the huge challenge of "knowledge integration" was within reach. My term (1982–88) as researcher in Lund was also reaching its end. Far from providing freedom to finish projects on hand, my calendar became filled with a host of new ones. Among these were invitations to participate in Sweden's contributions to the international program on Human Dimensions of Global Change (Buttimer 1988; see also Buttimer 1990); to deliver a special presidential plenary lecture at the annual meetings of the Association of American Geographers in Phoenix, Arizona (April 1988); to prepare a formal public lecture on an occasion marking the 550th anniversary of a *studium generale* that was allegedly established in Lund in the year 1438; to work toward the reinstatement of an IGU Commission on the History of Geographical Thought at the International Congress in Sydney (August 1988); and also to prepare for a lecture visit to the University of Ottawa (November 1988) with a view to consider their invitation to spend a term there as visiting professor in 1989.

The year 1988 was indeed busy and productive. Most delightful of all was an exploration into medieval *studia generalia*—Dominican and Franciscan—data for which led me to archival sources in Ireland (Buttimer 1989). *Geographers of Norden,* a collection of autobiographical essays and interviews with retired Scandinavian geographers, was also completed (Buttimer and Hägerstrand 1988). The Sydney IGC was a most enjoyable affair and our efforts to have the group on History of Geographical Thought reinstated as commission were successful with Keiichi Takeuchi as chair and me as secretary. A pilot project on the human uses of woodland was also approved; Stefan Andersson, Karin Hammersmith, and Klas Sandell were already at work on this as I departed for Ottawa in January 1989 (see Buttimer 1992a).

The years spent at Lund were filled with challenge, much of this emanating from joint initiatives with Torsten Hägerstrand. It was indeed a privilege to contribute to these diverse projects, symposia, and events. Research funding was generous and contributions to international scholarly activity were encouraged. The diligence, graciousness, and competence of Suzanne Krüger and Gunborg Bengtsson, and the cordiality of other occupants of *Djurhuset* during those years remain an indelible memory. Funding from the Wallenberg Foundation enabled us to establish a complete archive of tapes and transcripts from the Dialogue Project in Lund University Library Media center.[5] A second printing of *Practice of Geography* (Buttimer 1983c) did offer encouragement, and 13,500 copies of a

5. In 1996, a selection of these recordings with senior geographers was deposited in the library of the Royal Geographical Society in London (Buttimer 1996b).

Russian translation were in circulation. Yet I felt quite remote from the conversations among English-speaking geographers during the 1980s. The prospect of resuming a role as a geographer in a North American department in 1989 was therefore a welcome challenge even if it meant taking many pieces of unfinished writing with me.

Ottawa, Canada, 1989–1991

Ottawa. What a friendly place! By 1989 one was now in an electronic age—with PC, e-mail, and fax one could be in touch with the entire world from a hotel apartment in downtown Ottawa. It was indeed a tremendous boost in self-confidence to be a full-time participant in a geography department again. Colleagues were immensely cordial and students were enthusiastic. Socially, too, I was welcomed into several contexts beyond the university, interacting in French, English, Swedish, and even Irish. Bertram joined me in 1990 when he retired from his professorship in Lund and for him the snowscapes of the Gatineau were a ski-enthusiast's paradise.

Canada, like Sweden, generously promotes international exchange on issues of environment and development. Soon after my arrival the Swedish Ambassador, Ola Ullsten, introduced me to a number of "personalities" involved in environmental work, including those at the Royal Society of Canada who were launching the Canadian Program on Global Change. Students there also showed keen interest in ideas and I taught courses on "Nature, Space, and Time," "History of Geographical Thought," "Geography and Humanism," and a large undergraduate course on "Geography of Economic Systems" (see also Buttimer 1996c). In a small doctoral seminar, too, preliminary drafts of my work on root metaphors and knowledge integration were combed through by generous students who helped me to finish the text of *Geography and the Human Spirit* (1993). The Swedish part—the story on the making of geography in Sweden—which they found most illustrative, however, was excised by the publisher. From Ottawa it was easy to continue work for the IGU Commission, preparation for symposia in Hamburg (1989), Beijing (1990), Utrecht (1991) and eventually, Fredericksburg/ Washington (1992).

In addition to other unfinished manuscripts from the Dialogue project and other duties as IGU Commission secretary, the central research challenges of the Ottawa years sprang from the pilot project on human uses of woodland. Now with additional funding from both Sweden and Canada, a research exchange was orchestrated, culminating in a binational colloquium held at Ottawa in May 1990 (Buttimer,Van Buren, and Hudson-Rodd 1991).

Ireland Encore, 1991–

It was during a coffee break on May 10, 1990, that I found in my mailbox a copy of an announcement in the *Irish Times* about a vacancy for the professorship of geography at University College Dublin, applications deadline just a few days later. Immediately I picked up the telephone and requested application forms. The arrival of these application forms via the little fax machine at our home was Bertram's first inkling that another change of abode was in the air. In characteristic fashion he supported the idea. He has been warmly welcomed in Dublin as senior associate fellow of the Department of Mathematical Physics. Now and then, of course, he longs for the snowscapes of the Gatineau.

Dublin in the 1990s is a bustling and busy place. Student numbers in geography have virtually trebled since 1991 and the undergraduate curriculum is demanding and vibrant. Ireland rejoices in reasserting its place in European cultural life.Various programs enable students to study in different countries and Dublin has become a favorite destination. Cohorts of graduate students vary from year to year and funding is scarce.Yet there a growing diversity of research interests and field excursions are a treasured part of program. (See Buttimer 1996d and Buttimer and Pringle 1996 for a look at Irish Geography.)

Ireland in the 1990s sometimes seems like a foreign country. Rising consumerism, dramatically increased car numbers, massively increased investment in export-oriented production, mergers and takeovers, qualify this wee island as a tiger economy. Many corollaries—growing social inequalities, decimation of small and medium-sized farms and local services concomitant with the invasions of footloose commercialism—are worrisome. And the troubles persist in northern parts of this island despite gargantuan efforts from many sources within and without to find solutions.

My father's death in 1994 left a huge vacuum in the hearts of all my family. It seemed indeed appropriate that I should promote student interest in issues of environment and development—questions of sustainable rural life that had also been important for my father. Between 1993 and 1995 I coordinated a rather ambitious research project, funded by EU, with partner teams in Germany, Ireland, the Netherlands, and Sweden. Entitled "Landscape and Life: Appropriate Scales for Sustainable Development," this project endeavored to derive lessons from experiences of the 1950–90 period in these different settings (Buttimer 1992b, 1995b, 1995c, 1998b, 1998c; Buttimer and McGuaran 1994; Buttimer and Stol 1997).

It is good to be home, to give and receive and to sometimes feel that one is making some contribution. In addition to teaching and administrative tasks, there are endless boards and committees where duty calls. Many projects have actually been completed since 1991, despite a crowded calendar. International activity has even increased. I continued as secretary of the IGU Commission on History of Geographic Thought until 1994, when I assumed the chair after the resignation of Keiichi Takeuchi. In 1993, I was elected member of Academia Europaea and, in 1996, councillor of the Royal Geographical Society. It was a big thrill in 1996 when I was elected vice president of the International Geographical Union—a first for Ireland. This has entailed travel to distant places like Berne, Seoul, Pretoria, Stockholm, and nearer places such as Paris and Lisbon (see Buttimer 1995a, 1996e, 1996f, 1997). My hope is that in this role I may continue to further the ideal of international dialogue—surely one vital contribution that geography can offer humanity.

Reminiscence sometimes evokes nostalgia, sometimes remorse over unfulfilled promises and dreams; often, too, a tendency to view the past through the proverbial "rose-tinted spectacles." For me, however, the overwhelming sense is one of gratitude. How fortunate to have been a teenager in the 1950s in a rural Ireland then awakening to fresh possibilities, and then to have lived in the early 1960s in the Pacific Northwest, with its vast expanses of evergreen

and springlike blossoming of ecumenism and regional development. What a privilege, too, it was in the mid-1960s to be in Belgium and France where academic halls were bristling with radical
ideas of various hue. And Glasgow at decade's end was truly a joyful
time. For the early 1970s few departments could rival Clark's Graduate School of Geography—at least so we all convinced ourselves.
Lund afforded the attraction of opposites and the thrill of dialogue
between humanist and social engineering perspectives on society
and space. The 1980s in Sweden provided ample opportunity for
genuine cross-disciplinary and cross-cultural exchange at a time
when colleagues elsewhere seemed to become much more narrowly focused, some even hostile to the quest for broader themes or
mutual understanding (see Buttimer 1987b). And the brief *séjour* in
Ottawa brought that special gift of restored self-confidence which
survives through the turbulent 1990s. What the 1990s have actually
afforded are opportunities to revisit all these places—Seattle, Leuven, Glasgow, Paris, Lund, and Ottawa—each occasion delivering its
own "geography lesson": places change, people change, but certain
values retain their appeal—graciousness, courage, and hope through
the darkness that clouds the final years of a tired millenium.

I am grateful for all that is positive in my present life situation, and
for the myriad ways in which the past lives on in the present. Many
diverse worlds indeed remain alive as friends, colleagues, and students remain in touch via e-mail, fax, and telephone. Just a few weeks
ago an e-mail message brought the invitation to accept an honorary
doctorate in philosophy from the University of Joensuu, Finland, together with Professor Pierre Bourdieu—so experiences of home,
reach, and journey continue to offer challenge and surprise.

Been There, Done That, What's Next?

Did Theory Smother My Discipline When I Wasn't Looking?

John Eyles

I seldom admit to being a geographer nowadays. I have a nagging suspicion that our discipline is in its last throes, terminally wounded and, like a prisoner on death row in Texas, doomed inevitably to die. In darker moments, I feel that geographers like me bear some of the responsibility for the subject's fate, abdicating the field to those who, in my view, have made it almost terminally irrelevant in a world where applied sciences flourish and the civilizing disciplines languish. I recognize that my world is partial—one of many possibilities. But there is a homogenization of practice in Western countries. What is happening—Ontario today is similar to what happened in Britain in the 1980s, in Alberta in the 1990s, to what will happen soon in British Columbia, and to what was always the way in most of the United States—the construction of a world that demands of all individuals and institutions: What can you do for me? How can you contribute to the bottom line? How can you manage with less? Why do we have so many courses and degrees in history, sociology, geography, literature and so few in biochemistry, bioethics, software engineering, manufacturing and business practice, and so on? Yet geography still in so many ways provides a practical training for those graduating into this world. It is, when practiced well, a synthetic, interdisciplinary way of looking at the world, providing skills in critical and strategic thinking, explanatory and statistical analysis, and computer and information system use, portrayal, and application. Why is such a practical discipline at the undergraduate level in such desperate straits?

For me, the reasons are both external and internal. Externally, I agree with Gunther S. Stent (1969) and John Horgan (1996) that the external environments in and on which disciplines work are bounded. Human anatomy, geography, and more arguably chemistry are seen as finite ventures, which then must repeat or take flights to their margins to continue to survive. In some cases, the margins become the core—biochemistry, pathology. Ours remains the land of a thousand dances! Some of the dances have their bases in our engagements with social theory—let us recall the admonition in a British education paper: "theory in geography is like a London bus (pre-privatization!). Don't worry if you miss one, there's always another one following." And I want to use this chapter to reflect on my own engagement with social theory and my partial view of the engagement of others to point to how I think geography has both developed and become disadvantaged. In sum and cynically, we have not chosen our allies well, often desiring to be on the leading edge, especially theoretically, wherever that might be—economics, psychology, marxism, humanism, structuralism, literature, cultural studies. Part of our demise, I submit, is brought about by addiction to theory chasing.

Welcome to the World:
An Initial Engagement with Theory

I came to geography through a love of places, inspired, like for many others, by National Geographic and large glossy books on world geography. I was intrigued by the variety in the British landscape and how history could be read from field patterns, buildings, cities, routeways and so on (see, for example, Hoskins 1955). But "history" is in some ways not the right word. It is people and their ways of living, thinking, and being that are inscribed in the landscape. It was this connectedness between people and places that made me a social geographer. It was not, then, the broad sweep of history that made a place vital, vivid, meaningful. I wrote in *Senses of Place* about my undergraduate days in London:

> The streets themselves were places to explore and to travel. An important part of being in London was being able to move around it by bus and tube but primarily on foot. In the evenings and nights, I used to walk miles, usually alone, not only around the center—the city and the West End—but also around the residential districts. The walks I did most—indeed the ones I remember most—were into the East End. I do not know what really attracted me back again and again to this area. Did I expect to see the spirit of Fu Man Chu? Or drink in the same pub as Reggie Kray? Or buy seafood from Tubby Isaacs himself? Was it the image of East London as a community with lots of spirit culled from Dixon of Dock Green and a cursory read of Young and Willmott? The East End more than any place I have known evoked a mass of associations. I think at the time I expected to witness examples of these evocations. I still cannot dissociate the East End—indefinitely defined—as a place of attached sentiments and values. It was in my walks and solitary visits to East End pubs that I first realized, somewhat uncomprehendingly in the beginning, that a place is about its people. (Eyles 1985, 15)

Examining the relations between people and their places is very much a "hobby" now. I remain intrigued by conflicting interpretations of the landscape and how different groups want to build in their distinctive values. Thus, for example, Berlin fascinates me with the ideologies of imperialism, Nazism, and the cold war struggling to redefine the spaces and places of the city (see, for example, Ladd 1997).

But although my roots as a social geographer may be seen in these poorly developed notions of people in places, it was not until I became a graduate student that I made my commitment to become a professional geographer and the engagements with theory began. I should emphasize that my commitment has always been loose and ill defined. Gould (1988) wrote eloquently about his being "called to geography." For him, it is a calling that gives considerable freedom as the geographic way of looking at places sets few bounds about subject matter. But there are also constraints in that geographic eclecticism can lead to superficiality. In my work as a professional geographer, I have accepted the freedoms but not the constraints. I have always been happier at the margins of disciplinary turf, remembering with some frustration the hours wasted in discussing whether there was such a thing as a geographical factor or whether for locations to be meaningful they had to be relative, defined not only in terms of intrinsic characteristics but also of other places. The margins that excited me most as a graduate student were sociology and to a lesser extent anthropology. This, it seemed, was where the action was—for those interested in people, in places, and how these "phenomena" might be described, understood, and explained.

My years as a graduate student were spent, although I did not know it at the time, in a paradigm shift from traditional, descriptive geography of places to the quantitative analysis of spatial patterns. My first formal engagement with theory was with small t theory in which the relationship(s) between sets of variables are specified, operationalized, tested, and accepted (or rejected). At the time, the quantitative revolution seemed sterile, reducing people to the status of their (census) attributes, groups and communities to aggregates, activities to transactions and institutions to givens. It was only much later that I realized that a specific, quite elegant view of scientific activity was present, hidden in the mechanistic application of techniques—nearest neighbor analysis, factor analysis, simple regression and so on. Indeed, R. J. Johnston (1984, 31), in his recollections of spatial science concurs: "the main impression is one of only partial appreciation of the details of positivist arguments" with no coherent theory. Later, it would be claimed that the "quantitative reassertion in twentieth century geography . . . instantiated—using algebraic codes—a way of thinking that, in its most pernicious form, could represent qualitative issues about worth . . . to evacuate language of value judgement and to replace it with a formal calculus" (Livingstone 1992, 328; see also Taylor 1985). I am not so sure. But there were other influences at work in the academy if not solely in the classroom. My graduate experience coincided with Vietnam War protests, French challenges to the power to define higher education, Black power, the rediscovery of poverty in affluent nations, and the period of intense decolonization. The world was turning. The poor, the dispossessed, the outcast, the marginalized seemed poised to take their rightful places in the world. And that they would was "fact," as theory (big *T* this time—ideas on how the world operates) said so.

There were operating at this time, however, two related but eventually oppositional tendencies that have been labeled the liberal and radical critiques (see Johnston 1979). Richard Peet (1977, 242) sums up the differences:

> the starting point was the liberal political socio-scientific paradigm, based on the belief that routine problems can be solved, or at least significantly ameliorated, within the context of a modified capitalism. A corollary of this belief is the advocacy of pragmatism—better to be involved in partial solutions than in futile efforts at revolution. Radicalization in the political agenda involved, as its first step, rejecting the point of view that one more policy change, one more "new face," would make any difference.

The academy had to leave its ivory towers. Geography, like many other disciplines, was galvanized by the sufferings of the disadvantaged and oppressed. It talked less of differences and more of inequities. In my initial research as a university professor, I was much persuaded by the liberal view. My first published paper looked at behavioral patterns in east London (Eyles 1971). In it I argued that position in the social structure shaped access to social resources such as power and wealth that affected position in the spatial structure and access to spatial resources such as housing, shopping, and recreational opportunities. This access determined whether preferences could be pursued (or revealed) or denied (or repressed). This research was influenced by the behavioral turn in geography, with its emphasis on choice and action-spaces as well as the system constraints that are reflected in social position. The emphasis on resource allocation—a topic to stay with me in various guises throughout my career—was influenced by readings in the sociology of welfare and income distribution (for example, Titmuss 1962, 1965; Marshall 1965) and their geographical expression in David Harvey's (1971, 1972) essays on income redistribution and social justice in spatial systems. This reading had been "forced" upon me by teaching responsibilities. I had been hired at Queen Mary College (QMC), University of London, to lecture in social analysis, which meant I had a very practical reason to read widely in social theory as well as on the empirical topics such as class and stratification, power, poverty, and race.

It seemed at the time that I became more and more drawn to the radical perspective. Bob Colenutt and I, for example, tried to edit a collection on such approaches reflecting North American and European experience. The project failed. In the early 1970s there were still few empirical studies in the area. Further, Colenutt courageously decided to practice what he preached and left his academic position in Bristol to engage in community work in the working-class district of Southwark, south London. This was a shock. I can recall, too, reading Harvey's (1974) paper on involvement in public policy in which he claimed that there was no need for any further investigation of people's inhumanity to others—the point now was to change the world. Indeed, Harvey (1972) had set out earlier the "heroes and villains" of the theoretical world, when he argued that there were three kinds of theory:

• status quo theory, grounded in the reality it seeks to portray and only capable of perpetuating the status quo.

• counter-revolutionary theory, which obscures and obfuscates our ability to comprehend reality and the purpose of which is to divert attention to nonissues and toward nondecisions.

• revolutionary theory, firmly grounded in reality, dialectically formulated and able to encompass conflict and contradiction to offer real choices in existing situations.

It seems then that liberal perspectives were counter-revolutionary, obscuring the need for radical change. Yet I remained drawn to a neoliberal position that recognized the complexity of the social world, such that poverty must be seen as more than a lack of income but as a manifestation of inequalities derived from complex power relations (Eyles 1974) and that spatial inequalities—although they did not really interest me at the time—could not be tackled by spatial policies alone (see Coates Johnston, and Knox 1977). The marxist view of social relations seemed too simple for complex socioindustrial systems like Britain and the U.S. (see Giddens 1973). The mid-1970s were the age of mechanistic marxism, much influenced by Althusser, Poulantzas, and Castells. The elegance and symmetry of the postulates seemed in inverse proportion to their relations to the real world.

I admit, however, that I find their theoretical elegance appealing. I still do, so much as to wonder whether I am not a closet structuralist. In the 1970s I read and talked with urban sociologists and geographers (see Pickvance 1976) and while wanting to believe, found the tenets wanting. I remember the embarrassment I felt when Brian Goodey asked me if I was a marxist. The true answer was no, but I mumbled. Although outraged at the time, as an old fart in the discipline now, I probably concur with Brian Berry (1972, 77): "the majority of the new revolutionaries, it seems are essentially 'white Liberals,' quick to lament the supposed ills of society and to wear their bleeding hearts like emblems or old school ties—and quicker to avoid the hard work that diagnosis and action demand." He was being critical of both liberals and radicals. At that time, I saw that geography's focus should be on diagnosis, focused on how resources are allocated and on the role of power in such allocations. "Poverty and the distribution of real income in a spatial system cannot be understood without reference to power and inequality" (Eyles 1974, 74), which cannot themselves be understood without theoretical exegesis. I began to feel that conflict and/or constraints theory would be useful. The neo-Weberian positions adopted by Ray Pahl (1970, 1971) in his work on the city as well as that of John Rex and Robert Moore (1967) on the struggle for housing space in the racially mixed, largely working-class neighborhood of Sparkbrook, Birmingham, seemed useful, theorizing the bases for the allocation of scarce resources. I did not, however, follow this work up empirically. I discovered the content area in which I was to spend the rest of my career—health and health care—and with this a return to earlier engaged theories: marxist structuralism and humanized positivism. Were these retreats? I also began my forays into policy commentary. Was a different career beckoning?

If teaching social analysis as an undergraduate course and close reading of Harvey and the sociologists of welfare had shaped the first years of geographical practice, the next part was shaped by a closer connection with the East End in which QMC is located and by the development of a professional relationship to last ten years or more with David Smith, appointed professor of geography at

QMC. Smith is a quiet, private man, and although he was (is) always willing to engage in debate when asked, I gained as much from reading his work as from talking with him. In his written work, Smith (1973, 1977) was very much the diagnostic scientist, applying critical reasoning and quantitative methods to a subset of social problems. His interests in welfare and social well-being dovetailed nicely with my own on poverty as inequality and the structural parameters of inequality. But at QMC, the main theoretical discussion took place between David Smith and Roger Lee, between welfare and marxist economics. For me, Smith pointed up the relevance of geography and it seemed a minor, disappointing role—a descriptive one, emphasizing regional or place differences, albeit on a set of socially relevant variables. It was to be much later that I was struck by the fact that this is *the* role for geography—not a tremendously glamorous or leading-edge one but one that could impart analytic and critical skills to students through the study of what is familiar and often interesting—places and the people in them. The relevance of the discipline continues despite its boundedness and lack of significant contribution to the academic or practical worlds.

During the second half of the 1970s at QMC, I was also drawn closer to the issues and patterns of the East End. I have always chosen to study my own "backyard." I had begun to provide commentaries (in the spirit of the liberal academic!) on the many reports that were being produced on redeveloping and reconstructing the East End, especially with the closure of the docks. I began to become involved with action groups in the East End that were trying to push for more services. I also commented on spatial or areal policies in general, agreeing with the general thrust of the criticism of these initiatives, namely, that they were partial solutions to deep-rooted problems and would not in the long term make much difference to people's lives in working-class neighborhoods. I blush at such arrogance now and would probably argue that every little can help—the ultimate liberal pragmatist! This incursion into policy commentary did not mean any new theoretical developments for me. I soon left that field to those better qualified to comment, returning to it to update my review in the mid-1980s (Eyles 1987a) with a continued social justice agenda but no stated position on how to achieve change. I was still very much an academic commentator.

The East End and David Smith would combine to help me discover and define my content area of health care. David Smith, a
Ph.D. student, Kevin Woods, and I tried to interest both national and
local bodies to contribute to funding an institute for health research.
What we obtained was some contract research and the addition of
some other very good Ph.D. students. The middle years of the first
Thatcher government meant that there were very few public funds
available. At that time, I was not particularly worried by that lack. I
had always been a solitary scholar, willing to discuss, be influenced,
and share ideas but ultimately wanting to work on my own. This
solitariness may be a reason why so much of my earlier writing is in
the form of codifications—in texts—or (critical) review—in articles. This trait was exacerbated by my being on the margins of geography in my position as a lecturer in social analysis. But all these
influences and styles of practice came together in health and health
care, especially as I began to engage in an academic activity that
gives me great joy—using theories in specific content domains. This
I began to do as I immersed myself in the study of health care. I recognized, following Vincente Navarro (1978) and Lesley Doyal
(1979), that medicine was a powerful institution in capitalist society.
Medicine had subverted health by treating it like a commodity, with
the patient as an injured/diseased/disabled machine and the doctor
and hospital as technological, mechanistic repair shops (Eyles and
Woods 1983). In these ways, medicine becomes an apparatus or
structure that operates on but beyond the control of individuals. Its
importance, like that of law, is that it defines individuals and their
cultures in ways that resonate with the dominant orderings in the
economy. Medicine thus performs important functions for the reproduction of the relations of production, although challenges to its
(and others') dominance must be recognized. In this analysis, theory
(big T theory) was extremely important, and although my interests
and career would take other turns, theory would be played out again
in my examining of housing advertisements as signs (how dominant
ideas and their structures are inscribed in a particular landscape) and
in the iconology of Hamilton (Eyles 1987b; Eyles and Peace 1990).
Structural determinants would also appear in health policy writings
on the inverse interest law—that dominant, economic and political
interests determine what is recognized as a problem and is treated as

such, in the research of my Ph.D. students, particularly that of Martin Powell on territorial social justice in the delivery of primary care in London and in my first successful research grant application to establish the bases of resource allocation strategies in health care in England and to examine their consequences. Working with Jenny Donovan, I found the consequences to be engaging and subverting—my research and theoretical interests took a different turn.

Qualitative Methods and the Theories of Meaning

If the first phase of my career covers the years 1970 to 1983, the second starts in about 1980 and runs to about 1993. There are essentially three strands to this second phase which, to anticipate, led to my advocacy and use of qualitative/interpretive research designs and to the emergence of an individualized or person-oriented theory to inform much of my research, namely, interactionism-constructionism.

I have always been an eclectic reader. As I became dissatisfied with structural marxism's treatment of individuals and their experience, I began reading widely in social theory/empirical sociology/social philosophy. It is a commonplace observation for most that we are not coerced to perform our everyday routines. Further, our experiences can be the bases of collective action and resistance. Ideologies are not monolithic but are fractured, often contingent affairs. Yet they are obviously influential and do, at particular times and in particular ways, serve certain interests. For me, a nuanced view of the social system and its embedded social relationships, institutions, and experiences came from reading Gramsci and his various commentators. His notions of hegemony and resistance were well articulated in the research undertaken by the Centre for Contemporary Cultural Studies at the University of Birmingham, which was directed at this time by Stuart Hall. Their work encompassed working-class culture, black resistance, and the roles of education. Although the significance of the power to define roles, institutions, and so on would be enlarged by the influence of Foucault, I was taken by Goran Therborn's (1980) ideas about ideology and subjection-qualification. Dominant interests have the ability to define what exists, what is good, and what is possible but not for all time and not without resistance, as the historical scholarship of E. P.

Thompson (1968) on the making of English working class and Christopher Hill (1971) on the English Civil War period attests. In the resistance of subordinate groups, new ideas and experience-based values are formed. Barbara E. Smith's (1981) study of black lung in Appalachia demonstrates well the changing power relations between the mining companies and their workers as experience (and medical knowledge) altered perceptions, values, and the bases of resistance. These notions of resistance have been taken up in post-modern geographies of the Other. Popular consciousness as a basis for resistance has, for many major social groupings, been subverted by dominant economic and political interests. The inheritance of Thatcher-and-Reaganism is that moral ideology has been substituted for morality, and self-esteem and dignity for the individual have come at the expense of social solidarity. Classes and other significant social groups in Western nations now exist in harmony, not tension or contradiction (Frank Parkin's [1979] categories), and they oppose the Outsider (Japanese economic power, Islamic fundamentalists, Iraqi "pirates," emerging infections, aliens, and others "not like us"). These readings on the lived experience of the oppressed or disadvantaged were used in my theoretical development to emphasize the salience of contingency, fractured, negotiated meanings. And they also led, as they did many others, to an appreciation of the significance of human agency in relation to structure(s).

These readings informed the two other strands of this career phase, namely, my continuing research in health care and my own Ph.D. work on senses of place. In health research, my first funded project had, in the application phase, a policy focus. The project had been proposed in the light of the influential British Department of Health resource-allocation working party, which had established criteria for creating equal opportunity of access to health care for people at equal risk. The criteria were meant to be responsive to the relative needs of populations. In the end, the allocation exercise became largely a technically driven process of cost containment in which some health regions "gained" and others "lost" resources. Let me note in passing that these criteria were quantitative assessments of need, weighted and applied to populations—a technical exercise in engineering social justice, an approach that would revisit my career later (see below). But the 1984 funded project was meant to explore regional variations in perceptions and experiences of health and health care to comment on the utility of the region as an allocation device as well as on the inequalities in health status determined by the Black Report (Townsend and Davidson 1982). The research concentrated on those most likely to suffer poor health—the working class—and therefore more likely to benefit from greater access to care.

This research did not of course occur in an institutional vacuum. As reported earlier, QMC had successfully created a health research group that had attracted Ph.D. students from a variety of undergraduate disciplines—sociology, human biology, social policy, and geography. Two of those projects were nearing completion as the research was started. Jocelyn Cornwell (1984) had submitted her accounts of health and illness in Bethal Green, East London. "Hard-earned Lives" documents the public and private accounts of health and illness in this working-class population. For my purposes here, Cornwell adopted the work of interpretive medical sociology as this was conducive to validating the patient's point of view. It is an approach that leads readily to serious treatment of how people "actively and creatively produce and reproduce the meaning that sustain their social world in every moment of their interactions with other people" (Cornwell 1984, 19). Jenny Donovan (1986) was completing her study of health and illness in the lives of black people in London. She too adopted an interpretive approach—the social world being seen as intrinsically meaningful being created, recreated in every social encounter (Silverman 1972; Giddens 1976). In our work on working-class health in England, we adopted Alfred Schütz's sociological phenomenology. Schütz (1972) regarded the aim of the social sciences as being to interpret the subjective meaning of social actions through understanding the stock of knowledge, recipes for action, held and revised by individuals. This stock contains ideal types that facilitate communication and that are indexical, acting as markers for sets of information and meanings. Scientists can discover these recipes, stocks, and meanings by creating second-order constructs or typifications based on the experiences and meanings of the individuals of interest. Donovan and I utilized these ideas to examine the definitions of health, sickness, and care in the west Midlands (Eyles and Donovan 1986).

Common to all these projects, then, is a commitment to inter-subjective understanding through theories of meaning and to a set of methods appropriate to the theories and their related research questions. Arguably, the interpretive approach is a methodological stance, which for geography borrowed heavily from anthropology. It is perhaps at this stage that I lost interest in geography and have periodically tried to regain it ever since. For to understand the lived experiences and meanings of individuals the researcher had to explore, theme, and describe fully that which was reported and observed. Clifford Geertz's (1973, 10) term "thick description" is often used, this entailing:

> What the ethnographer is in fact faced with—except when (as, of course, he must do) he is pursuing the more automatized routines of data collection—is a multiplicity of complex conceptual structures, many of them superimposed upon or knotted into one another, which he must contrive somehow first to grasp and then to render. . . . Doing ethnography is like trying to read (in the sense of "construct a reading of") a manuscript—foreign, faded, full of ellipses, incoherences, suspicious emendations, and tendentious commentaries, but written not in conventionalized graphs of sound but in transient examples of shaded behavior.

I became increasingly involved with method (Eyles and Smith 1988). But the purpose here is to link this ethnographic description with theory (and with the third strand in this career phase). The lack of such interpretive design may be rephrased to "grasp the dynamism of the lifeworld" (Buttimer 1976) and related to humanistic geography.

The connection between interpretive philosophy and humanistic geography was made explicit by David Ley (1977, 504–5):

> As social geography follows its agenda and dips beneath spatial facts and the unambiguous objectivity of the map, it encounters the same group-centered world of events, relates and places infused with meaning and often ambiguity. Husserl, in his later writing, characterized this realm as the lifeworld. More recent philosophies like Schütz and Merleau-Ponty have urged that this reality encompassing most experience is not irrational and impossible to study.

The purpose of humanistic geography was then to study the most distinctively "human" aspect of people—meanings, values, goals, purposes (Entrikin 1976). It was largely an explicit reaction against the dehumanizing tendencies seen in spatial science (see Cloke Philo, and Sadler 1991). But I recall the struggle also being between humanistic and marxist geography with the lines drawn between the relative power of human agency and structural forces. In some ways, this battle was a nonissue before it was fully played out as Giddens's ideas on structuration, incorporating the equally valuable integrating work on volition and constraints in time-geography, and became the New Theory.

My own theoretical interests within the humanistic tradition were somewhat different, although I too have made appeal to the interpretive community of structuration (see Lamont 1984 on the power of such a community). In reading humanistic geography, I was struck by its remoteness from the world. It provided intelligent commentary but, with some notable exceptions—Ley 1974 and Western 1981—it failed to engage the lived experiences of those whose meanings we seek to capture. I decided to investigate humanistic geography empirically, focusing on senses of place (Eyles 1985). This investigation of a small town in the English Midlands (Towcester) utilized survey and interpretive research designs. It adopted a Weberian-Schützian framework for interpretation, producing second-order constructs to represent senses of place. These ideas were taken up in and as part of the Ph.D. work of David Butz in Shimshal, Pakistan. Using Habermas's theory of communicative action, Butz examined and analyzed key features of the changing lifeworlds of Shimshalis such as the significance of portering (Butz 1995) and the role of pastoral resources (Butz 1996). Later we combined to reconceptualize senses of place by bringing together reconsiderations of Towcester and Shimshal (Butz and Eyles 1997). For this, Thomas Ingold's (1992) reworkings of environmental psychology were incorporated and the social, ideological, and ecological dimensions of sense of place were fully realized. In the joint work, two theoretical conclusions were put forward. First, sense of place is constituted through social and material circumstances but as an ideological effect it can influence the reproduction and transformation of those circumstances. Second, ecology is fully integrated into an attribute of sense of place such that it is an emplaced aspect of the lifeworld. These attempts to theorize ecology or environment derive directly from Butz's work. The concepts do, however, loom large in the third phase of my career in which the interface with theory is perhaps less direct than during this second phase. But as may come across to the reader, the engagement with humanistic geography had been a difficult one, reinforcing my view of the marginality of geography, or more positively, its merging with other sciences. The chorus sung by humanists (and others) that place matters is for me largely limited, special pleading; so too do family, prices, values, and interests. No such statement is particularly profound or useful on its own.

Small *t* Theory and Science and Politics: The Engagement with Research Transfer

This third phase runs from around 1990 to the present. It has taken a long time but I think that it is in this phase that I found out who I am theoretically. This period began with my leaving QMC, now Queen Mary and Westfield College, and arriving at McMaster University, with its somewhat complacent, elitist view of itself and its research significance. With its innovative health sciences center, into which I was immediately plugged, it has focused on problem-based learning in education, knowledge production and transfer, and the importance of teams in carrying out complex research tasks. I plunged into this pool of research, taking advantage of the opportunities the environment presented to work with a variety of teams. I have had the good fortune to work with sociologists, anthropologists, doctors, nurses, toxicologists, epidemiologists, chemists, geologists, biologists, philosophers, psychologists, policy analysts, psychiatrists, economists, and geographers. But in this eclecticism, I could see some obvious common threads—perhaps the most important of which theoretically was my seeing interactionism and constructionism as subtle ways of understanding meanings. As usual, I was a slow learner. Peter Jackson and Susan Smith (1984) had already brought this to the fore in their wonderful book on exploring social geography. They cite Paul Rock (1979, 94–95), who sees it developing from "the conjunction of pragmatist epistemology, the journalistic urgings of Robert Park and the efflorescence of new and unknown terrain." My influences in this regard were largely George H. Mead (1934) and Herbert Blumer (1969). Thus human beings act toward things on the basis of the meanings that these things have for them; the meanings are the product of social interaction; and they are also modified and dealt with through an interpretive process that each person uses to deal with situations. There are obvious linkages to Schütz's recipes for action and to the interpretive tradition in general. At root, then, social life is seen as socially produced or constructed by the actors who participate in it. There certainly remain the problems of dealing with order and structure in interactionism but it helped make sense of the research situations with which I was dealing.

Upon arriving in Canada, I extended the lay perceptions of health and illness work I had undertaken at QMC. Working with Andrea Litva, I examined middle-class health issues in Ontario and later working-class ones in the rural parts of the same province using an interactionist perspective. It allowed an exploration of contingency and negotiation in how people deal with illness episodes or the health care system (Litva and Eyles 1994). I also used the same framework with Jamie Baxter, Tim Sly, and Martin Taylor on the meanings of risk and hazard in low-exposure situations, that is, in environments that nonresidents would often consider benign. Baxter and Taylor were also part of the team that investigated the effects of the Hagersville tire fire in 1990, demonstrating that the explication of meanings, aided by interactionism, can be utilized in contract research and for research transfer purposes (Eyles and Sider et al. 1993). In these projects, the problem of order has not been ignored. It has sometimes been presented as context or history, as in our studies about anxiety over waste (Eyles and Taylor et al. 1993). For our studies of risk and hazard, we have utilized the notions of risk society developed by Ulrich Beck (1992) and Giddens (1991) which postulate that late modern society is qualitatively different from previous societies in that it presents the potential for technological catastrophe, with complex, expert systems possessing the capacity to fail and harm (Perrow 1984). On other occasions, order has been specified in statistical terms, largely through the increased use of combined qualitative and quantitative research designs (e.g. Walters et al. 1997; Taylor et al. 1997).

In fact quantitative designs have been very important in this third phase of my career as they have enabled me to think again about earlier theoretical issues, especially those relating to small t theory and the relationships between resource allocation and social justice. One of my earliest collaborators at McMaster was (and remains) Stephen Birch, a British-born and -trained health economist. Being thrown together by the McMaster system, we immediately identified similar research interests (and senses of humor). We had in fact both retained an interest in the British debates and policies concerning resource allocation. In the context of contract research, we were asked to establish the "fair" share of provincial health resources for a northern Ontario community that wished to develop a health service organization model of care delivery (i.e., all health and allied services from one location and based on one budget—a departure from the usual physician-only provision on a fee-for-service basis). This project was largely a technical exercise (Eyles et al. 1991), although we began to see the importance of defining "need" and criteria for assessing the efficiency, effectiveness, and equity in the distribution of resources. In the latest manifestations of this work, we have assessed the performance of the Canadian health system over time (Eyles, Birch, and Newbold 1995) and have begun to work on the effects that population heterogeneity has on the distribution of health status and eventually the distribution of resources. Underpinning all these works is a vision of a socially just distribution not dissimilar to that which engaged me in the 1970s—but now perhaps better articulated—that there should be equitable access to scarce resources (e.g., health care, the social determinants of health) and that the researcher's role is largely diagnostic, with suggested prescriptions for incremental change: a liberal stance with a commitment to a neo-functionalist perspective! But more on structuralism later.

During my years at McMaster, one of my most fruitful research partnerships has been with Martin Taylor in geography. We have worked together particularly on the psychosocial and health impacts of exposure to solid waste facilities—incinerators and landfills in particular. Taylor, a superb quantitative researcher and fine social psychologist (see his work on mental health [Dear and Taylor 1982] or noise [Taylor 1984]), invited me to join the team as a qualitative methodologist. I did, and I feel that I have made a methodological and theoretical contribution to the group's work (see above). What Taylor did for me was rekindle my interest in scientific problem-solving, dormant since my master's degree in the late 1960s. Careful variable specification and operationalization and attention to analytic strategy and techniques became central concerns. The role of theory in such designs is important in the sense of specifying the relationships between variables, often in the form of environmental stimulus—psychological response (see the stress-and-coping theories of Lazarus and Folkman [1984]) or contaminant exposure—disease or health outcome (see the biological plausibility ideas used in epidemiology—Susser 1988). These are small t theories but extremely powerful ones in that they allow for strong associations to be claimed at the individual level, vital for attributing risk to particular hazards. Small t theory is helpful in allowing a recognition that whereas the biophysical world (reality) is given meaning through social interactions, it has a concrete existence beyond the construction of individuals.

This relation between the biophysical world and its meanings has been given further impetus in my research by an interest in the connections between science and politics and the role of evidence in decision making. This interest has come about through my holding a chair in environmental health, awarded by the three granting councils in Canada, in 1993. The chair has been the focus of much of my research activity since 1994. The group, whose core is Donald Cole, a physician-epidemiologist, Susan Elliott, a health geographer, Michael Jerrett, an environment and health modeler, and many graduate students and postdoctoral fellows, has focused on several strands of research—health and environmental quality; risk perception, assessment, and management; and environmental health policy analysis and evaluation—but little explicit geography! A key theme has been on meanings—of lay publics in risk situations and of science and policy in the use of evidence and decision making. Science is seen as a socially constituted and designed practice, in which claim making is particularistic rather than universal (see Hannigan 1995):

a theoretical stance that has its antecedents in frame analysis and social constructionism (see Ibarra and Kitsuse 1993; Gamson and Modigliari 1989). This work dovetails theoretically with the interactionist perspective for the study of individual responses. We have examined the construction of science and policy claims around ozone depletion, ultra-violet radiation, and skin cancer (Garvin and Eyles 1997) and the rhetoric underlying the health-for-all policy strategy (Iannantuono and Eyles 1997). What is prefigured is more research activity around the role of evidence (and stories) in policy making and the use of evidence in, among other things, the evaluation of the effectiveness of policy instruments. In this regard, research transfer and working directly with the users of research evidence has become a significant, explicitly atheoretical stance in, for example, the areas of national and provincial health policy and the Great Lakes and Ontario air quality. Underpinning these projects are interactionism and constructionism.

Yet these linkages beckon me to untried activities—working with the corporate sector around risk management and evidence-based decision making. In fact in the resource-strapped environment of Ontario universities, this sector looms large and potentially larger. Indeed many of the national research competitions in Canada demand private-sector partners. In my first interactions with these players, I felt great discomfort, especially as health outcomes are often seen as one of the (adverse) implications of much corporate activity. Further, their agendas are very different from those of my usual research subjects, even when in the role of good corporate citizen—with a more restricted view of an issue and its context than that usually present in the academy. "Why should we invest in this?" "What's in it for us?" "How will we look?" are questions often asked. Now it may be that the corporate sector merely asks bluntly the questions that most pose around any research partnership. Yet usually in the corporate sector, a limited, often less critical, range of endeavors seems to be permitted than elsewhere. So, in confronting this research audience and these potential partners, I am a scientist with specific technical and problem-solving skills. To return to my theme in this chapter, geography is largely absent and irrelevant. But I still believe that I would not be doing what I am without the training geography gave me.

Conclusions

I see the current phase of my career ending, formally in 2001, when I give up the directorship of McMaster's Institute of Environment and Health. So if history repeats, the next iteration will begin in 1999—a leave year! In this phase, I see a role for structuralism because I want to think about the role of macrosocial theory in helping our understanding of environmental health issues, specifically the roles of knowledge, institutions and civil society. I also see increasing attention being paid to the role of evidence in science and policy. I shall also extend the work on meaning at the individual level again in terms of the take-up of information. How do people process information? What cognitive models do they use? There are interesting developments in psychology and anthropology to guide my thinking. In most of this, the production and interpretation of evidence and the construction of action-plans and policies, interactionism-constructionism seems likely to remain a plausible and useful social theory.

I have tried to capture my engagements with theory from early stumbling beginnings to a later certainty that theories (and methods) are dependent on the research questions asked. Theories are not ideologies. They are, as Arthur L. Stinchcombe (1968, 3) says, useful "to create the capacity to invent explanations." Herein lies my discomfort with totalizing theories such as marxism and some forms of structuralism. In their cruder forms, they purport to know the world before it is investigated. My own use of theory has been quite eclectic, possibly opportunistic, although I have tried to review the significance of theories in general for the research activities of my context areas (Litva and Eyles 1995; Eyles 1997). But, by and large, two major stances may be identified—small *t* theory to examine associations, underpinned by a normative commitment to social justice through accessibility and opportunity for all, and interactionism-constructionism.

In this eclectic journey I have been largely silent about geography, since perhaps the mid-1980s, although I remain a member of the health geography community. At the danger of sounding like the Berry of the 1990s, I must say the incredibly nuanced theoretical and philosophical debates, the frequent lack of attention to methodological rigor, and the liberal borrowings from other disciplines have left me feeling that geography is largely irrelevant and that the world has passed it by. I suppose I do not feel the discipline's research questions are always clearly articulated in a programmatic way. Further, the engagement with literature and the increasing attention to the cultures of others are praiseworthy, important activities. So too is the concern with the history of our own discipline. But, in my albeit marginalized view, these concerns seem to dominate what is our leading edge. But then I am reminded of how many geographers there are practicing their skills learned in geography departments in nonacademic settings. The evidence is largely anecdotal but compelling. But I hypothesize that the skills they use are primarily analytic and quantitative techniques, GIS, critical appraisal and synthesis, and a systematic and strategic understanding of the world rather than literary critique or queer theory or body politics or ecology in sense of place. But my self-imposed marginalization may give me a partial view.

I have in the last fifteen years pursued what I regard as interesting "puzzles" increasingly in applied settings. I also want to do things I have not done before—research in accident settings (although "accidents" are seldom accidental), logistical regression, real-time, real-place evaluations—and this desire has taken me further from the core of geography and its rather monastic debates. So after close on a thirty-year engagement with theory, I ask, Is it irrelevant? No, it is not. Without theory, there can be no establishment of research questions, no informed understanding, and no attempt at explanation. And as I have shown it takes time to find out what theories can best answer the empirical questions you want to pose. Does theory's role mean that geography is dead? I am not sure. It lives very fitfully with its boundedness. Do I care? I think so—because of what geography can give its students. Do others care whether I care? Probably not. But what is happening to geography departments in North America, including my own—now the School of Geography and Geology—should make us reflect about our purposes and mission.

4 Through the Glass Darkly

Re-Collecting My Academic Life

Kevin Archer

Tampa, Florida

At the University of South Florida (USF), I teach the usual array of urban and social geography courses for which I am considered to have expertise. Being accustomed now to the academic context of the United States, I include in this teaching, in turn, quite a bit of missionary work for the discipline, which even my faculty colleagues have a difficult time placing in a university setting. As a vice-president at one of the institutions where I have been once put it—"geography has no place in a research institution as it is largely remedial, descriptive education." This missionary work I find particularly ironic, however institutionally necessary it may be, because I originally came to geography to loosen the disciplinary hold over my own research and teaching interests. And so I talk and write much about intellectual integration and the importance of a critically understood holistic approach to what is essentially a holistic reality (Archer 1995).

But significantly, my experience at USF, like my previous experience marginally hanging on at the University of Maine, has drummed into me the very importance of effective teaching, partic-

This chapter is dedicated to my very early mentors, Mark Kabush and Bob Brubaker, who opened the door; the late John Bradbury and Ruth Fincher, who pushed me farther through it; and finally, to Erik Swyngedouw, who continues to be with me, here, there and everywhere.

ularly at the undergraduate level. This is so not just because my normal teaching load may be higher than it would be in, say, a Ph.D. department at a more research-oriented institution. Rather, the focus on teaching at these institutions has made me more aware of the dynamics within the classroom which then better illuminate important aspects of my research outside class. This is also so not just because it is now obvious to me that I "reach" more people in the classroom than I otherwise would by my formal publications, which no one, apparently, has time to read anymore. It is more a result of my better understanding of the classroom setting as both a microcosmic "community" in the making and, yet, a community already oppressively structured according to traditional pedagogic standards. This very contradiction between an evolving, and ultimately quite bounded, community in turn helps me better understand the essential contradiction embedded in the exurban dreams of the developers and residents of the "new urban" places I now study, like Disney's town of Celebration outside of Orlando, Florida (Archer 1996, 1997).

The specific context of USF, of course, has helped me tread this path, with its large contingent of mostly urban, so-called nontraditional, and part-time working students from virtually all social classes. But I also think that most critical faculty have not put enough emphasis on the importance of the classroom as a laboratory setting within which theoretical perspectives can undergo concrete trial (but see hooks 1994; Giroux 1997). I know I came late to this notion, more than once considering classroom teaching as merely an obstacle to my real desire, research and publishing, as per my training and academic socialization. But now, it is this very singular desire to publish that seems so odd to me, and not only because of my present teaching load. After all, what led me down the path of an academic career was an experience in high school many years back that I eventually, it seems, had forgotten. Two enlightened teachers at the time, Mark Kabush and Bob Brubaker, had the intelligence and the fortitude to go against the grain of the institution to form an "alternative school" based on then evolving pedagogical notions of collaborative, active and participatory learning. The program they set up—"Counterbalance"—already included many of what I now call classroom decolonization techniques and characteristics which I have belatedly (re)understood as important for the kind of critical education I am seeking to facilitate today, both in geography and in the new undergraduate program I helped to create at USF, "Learning Communities." It was my involvement in this high-school program that turned me around at the time by turning me on to my own education and, indeed, potential role as a critical and active citizen of the world.

That's the funny thing about it. The higher academic training that I underwent after high school not only proved nothing like my alternative-school experience but actually biased me against the very things that originally led me down the academic path. I was trained in university to be an independent, original scholar, and to be considered original was, of course, to be as independent a thinker and writer as possible, solitarily questing for new knowledge. The classroom was considered merely the place where already known things were discussed, the individual laboratory or study office the place where new knowledge was generated. And it was clear which activity place was held in higher esteem along the increasingly monklike path toward the Ph.D.

Leuven, Belgium

That such higher academic training remains essentially Western and ultimately patriarchal, among other things, is now clear to me and not only because the monklike ideal is embedded and even reinforced in the faculty tenuring process even after the Ph.D. Also clear to me now are the adverse social implications of the Cartesian-like separation of mind-work from the rest of reality in the very institutionalization of the academy, as Sohn-Rethel (1978) argued so long ago. As I was going through the process, however, it all seemed so natural and, surely, something I successfully could accomplish, if only I put my mind to it. And I did eventually receive my various degrees from well-respected, research-oriented universities on the way toward establishing a proper academic career. So, what's up with my present talk about the significance of effective classroom teaching?

You see, I never truly understood. Theoretically, I was aware, but my awareness never was a real part of my life. The practice, as it were, was missing, even as the theory was getting deeper. It was not until the morning of March 11, 1987, that understanding for the need for practice began to dawn, just as our firstborn took his first breath. I had been shopping for announcements—the local custom—and it hit me, at least initially. As the printer put out the card: "Met grote vreugde melden wij u de geboorte van ons eerste kindje" (It is with great joy that we announce the birth of our first child), I began to think carefully, more carefully, surely, than the last few days leading up to the event. Here we were, my spouse, now Dr. Ingrid Bartsch, and I, in Leuven, Belgium, me on scholarship to do Ph.D. dissertation research and her with baby, just after her master's degree. Everything was planned and, yet, as I left the birth room at H. Hartziekenhuis at Naamsestraat 105 to perform my traditional fatherly duties, all this good planning seemed inadequate, singularly and nakedly theoretical. Two aspiring professionals now had another life to consider, to be sure, but there was more to it than that, for both of us. And I, it now seems, was the last one truly to understand.

So far, it has been over a decade in which this last realization of mine has evolved and somewhat matured. A long time of practice and, probably, a longer time necessarily to come, before all the details are embedded in me, if that is even possible. Before, Ingrid and I had met as graduate students in geography at McGill University, she being attracted from biology into wetlands biogeography and I being attracted from urban economics to a more realistic urban geography. The late John Bradbury and Ruth Fincher, both then at McGill, ensured that I chose a master's degree in geography over economics, even though my advisor in the latter discipline told me that geography for a master's degree would mean studying merely "ersatz economics" ("precisely why!" . . . I thought at the time). But still, even with Ruth and John's guidance I was not very aware of the margins of my universe, margins along which Ingrid was already treading as an emerging woman scientist (but that, like much of the rest of her life in connection to me, is for her autobiography to relate). I certainly did not consider myself much at the time. I was out there, white, man, hetero, quite "normal" to the world, even quite heroic, in my own mind. I was in a foreign city, on my own money, making my own decisions, leading my life with no apparent obstacles that could not be overcome. In short, my color, gender, and sexuality were invisible to me, because in practice these personal characteristics seemed so invisible to the world.

Of course, this all sounds so naïve, even trite in the current intellectual context. I am certainly aware of that. But what I find interesting is that my very "whiteness," "masculinity," and "hetero-sexuality" have only very recently been explored in much critical depth (e.g., Cornwall and Lindisfarne 1993; Pfeil 1995; Seidler 1995). All of these "personal" characteristics are as fluid as any oth-ers, any one fusion not that much like the next. Yet, this rendering visible of the hitherto invisible still cannot overcome the canonical social context that ensures that these sorts of differences are more normal than any others. And this is what I came face-to-face with: being part of a sedimented social canon the personal effects of which I really knew very little even though I seemed to know so much, theoretically.

Which brings me back to Leuven. To get to Belgium, I first left McGill to travel to Baltimore for a Ph.D. My decision to depart to work with David Harvey at the Johns Hopkins University did not seem much to me at the time. After all, I was a foreigner in Canada and was ready to come back across the border. The personal rub, at the time, was that Ingrid could not follow me because she was a Canadian citizen and we were not officially married. As I look back, I realize this should have been the beginning of my enlightenment. But it was not. The solution seemed so simple. And so Ingrid came down for (American) Thanksgiving week, we got married in the Baltimore City courthouse, and she went back to Montréal while her immigration papers were being prepared. In the meantime, I got to know my most important academic mentor better and, just by chance, got mixed up with some planners at the university, particu-larly John Dyckman, Erica Schoenberger, and a young Erik Swyn-gedouw: further, just by chance, I was asked to help set up a joint university–regional French government–funded center for plan-ning studies in Lille, France.

The opportunity to go to Lille was exciting for both Ingrid and me, she having grown up bilingual in Montréal and I being able to begin (again) practicing what French language skills I had acquired previously. But there was more. I met many a European scholar and was immediately immersed in a new public life, both professional and academic. And it was this experience that got me intellectually interested in the similar economic and social histories of the Flemish and the Québeçois, which would eventually lead us to Belgium. Yet, again, the invisibility of my status to me made it very difficult to understand the evolving problems within the newly fused personal/professional relationship. In effect, Ingrid, with all of her academic credentials and professional experience, was in the process of being almost completely privatized in Lille. Although there were some opportunities to use her language skills for translation and interpretation, these proved too sparse to bring much satisfaction.

Too sparse, in turn, was my own understanding of the situation. Even after Ingrid eventually took a job in Germany—meaning long absences from Lille—I was only beginning to paint the picture, yet still with overbroad strokes. Being called back to Baltimore for university residency requirements after about two years in Lille helped narrow the strokes, to be sure, but only to a limited extent. After all, Ingrid did eventually land a job with the State of Maryland and was able to find laboratory work, on and off, at Johns Hopkins. But, again, after the year in Baltimore we were back in Europe on my scholarship, to do the research in Belgium that I had dreamed up in Lille. And, again, Ingrid was privatized, linked to me on formal occasions but otherwise quite invisible publicly, regardless of her professional credentials and experience.

Which brings me back to the H. Hartziehenhuis, spring 1987. It was clear to both Ingrid and me that our firstborn's arrival was going to mean more than merely an addition to our household. While I was then only slowly beginning to understand Ingrid's frustration over her own previous privatization, she was quick to determine that the need to care for a child meant that some important career choices had to be made, rather fast. My scholarship to Belgium was soon to be finished but, more importantly, Ingrid's own academic career after her master's degree had been disrupted by her travels with me. Having a child seemed then to fuse our path even tighter necessitating decisions that, at the time, appeared to me to be not only increasingly out of my control but, indeed, somewhat compromising in terms of our earlier ideals.

But that was the point that I still missed at the time. I still assumed

I knew "our" ideals, perhaps mostly because both Ingrid and I had been academically trained in the same manner. Surely a research-oriented study-and-career path was our destiny, particularly now that we had so tightly fused ourselves together as a family along this path. Had I then been more perceptive, however, I would have been more prepared for what was really in store for us. After all, Ingrid already had "compromised" her own academic trajectory by joining me in mine. Although that certainly was the result of personal choice, the whole idea that this sort of choice can "compromise" one's choices and chances already speaks to the oppressive and ultimately anachronistic nature of the monklike existence demanded by our chosen academic career path. Yet, at the time, I could only glimpse this idea, however prodded by the birth of our baby in that small town in Flanders.

Bangor/Orono, Maine

That it was just a glimpse seems very strange to me now; or perhaps not. I use my own experience as an example in my classes today as a means to talk about the structuration—behind our backs and out of our sight—of our life experience. Being the norm within the social canon renders seeing the canon that much more difficult. And so it was, even as Ingrid and I discussed where we were to go after Belgium. I was then A.B.D. on my Ph.D. path and so potentially could relocate anywhere to finish what was supposed to be a dissertation on the evolution of ethnonationalism in Belgium. Because of this mobility, and the need to take our firstborn into account, we eventually decided to plan our career path more as tightly connected equals, with Ingrid now to determine where we would end up by choosing where she would begin her studies toward a Ph.D. She already had followed me, now I would follow her, and then she would follow me again, and so on.

But, as I now understand more fully, "following" each other was not exactly the right way of understanding the matter. Sure, it was Ingrid's ultimate choice to pursue a Ph.D. at the University of Maine, yet this "personal" decision was made within the context of what was best, of the choices she actually had to consider, for all of us in the family. The existence and extent of such things as stipends, housing, day care, and potential professional opportunities for her spouse, all played a role in Ingrid's final decision; things that surely narrowed significantly her range of choice. What I did not realize at the time was that this range of choice would get even more narrow the further we, as a family, deviated from the professional and institutional norms of our chosen career. I should have understood much sooner that the more Ingrid and I and children were fused packagelike in our career decision-making, the more difficult it would be to fit the traditional mold of our profession; the more difficult it would be, in other words, to be successful in both family and career; monklike we both, increasingly, were not able to be.

For me, the Maine experience was a singularly significant learning experience. There, I experienced a privatizing process similar to, if not as profound as, that Ingrid experienced following me around. I had no public role there to take up, except as an imperfectly known spouse and child caregiver during business hours and mysteriously unconnected library gnome nights and weekends. I slowly came to realize the similarities between my growing frustration at this role and that of Ingrid before me, something that I still believe marks the real beginning of my practical awareness of marginality, however by proxy and however incomplete this awareness may continue to be. I was, put differently, more and more conscious of myself and my role in the new social context and how I, in turn, was considered and ultimately treated by others, both inside and outside of institutional contexts.

Now, I know that I cannot make too much of this experience of marginality, nor do I intend to. But that it went beyond mere abstract theoretical voyeurism was particularly important for my own understanding of what Others have been asserting; I was no longer Kevin Archer, independent, educationally well molded, internationally experienced, public intellectual monad-on-the-make, but, rather, Kevin Archer, stay-at-home spouse of Ingrid, father of Kieran (and, eventually, Delaina), and otherwise private individual apparently going nowhere, rather quietly. This practice, no matter how superficial it may seem to Others, was very enlightening to me, if quite slowly and mostly after the fact. I understood myself increasingly as a related part of a related group, the relations of which fit very unevenly, if at all, within the canonical boundaries of the academic career path. This experience allowed me to decenter myself, if bit by bit, within my personal life-narrative while, at the same time, introduced me practically to some of the social and institutional vicissitudes lived by those on the canonical margins. (Again, I emphasize "introduced" here.)

The time at the University of Maine also brought home to me the relative marginality of my discipline within the United States. Although the university is a major land-grant institution and took part in the major NSF funding for a national GIS center consortium, it does not have a geography department. Although there were some geography courses taught within the anthropology department—largely thanks to Victor Konrad, a cultural geographer at Maine at the time who was there mostly to direct the university's Canadian Studies program—there was relatively little chance for me to attain a faculty position, even as an adjunct. This was certainly a final eye-opener for me. Even though I had been trained in other disciplines and thus felt competent to teach introductory courses in a number of social sciences, my Ph.D. degree in my chosen field simply was not considered adequate for this purpose. Thanks to Victor Konrad—and by chance my time at Canadian universities—I was eventually able to secure a professional position in the Canadian Studies program and, eventually, a postdoctoral fellowship at the new public policy institute formed during the time I was there. The former position included some opportunities to serve as an adjunct faculty in geography, mostly to spell Victor from teaching from time to time, and the latter position allowed for some teaching in the public administration department.

Neither position, however, really fit the kind of career path for which I thought I had been trained. Each position was temporary, somewhat out of my field (teaching the "Geography of Maine" after being there only one semester? "Cultural Geography"?), and surely not research oriented. Nevertheless, as part of a fused family group, I had to find some way to make an income, however off-task and out-of-personal-career-line the job might seem to be. But the lessons of Maine went further than that. I now understood better the characteristics of the realm of adjunct-hell and of otherwise hanging on and around universities for whatever, mostly off-task positions, might come available. This disjuncture turned out to be good

preparation for my eventual understanding of Ingrid's initial problems at USF as well as good, practical experience with the social implications of the growing use of temporary, adjunct, or otherwise
part-time, nonbenefited faculty on increasing numbers of university
campuses.

The experience at Maine also brought home to me the importance of effective teaching, particularly of students largely ill prepared for university and not likely to go further in education than
the bachelor's degree. The key, it increasingly seemed to me, was
somehow to facilitate active and critical engagement on the part
of such students with both the texts they were reading as well as
the world around them. Here is where I became first familiar with
what I now consider to be the main problem with the usual way
university teaching is undertaken: most classrooms essentially are
colonized by professors whose control over colonials remains all-
powerful and oppressive, hardly yielding of critical, active agency on
the part of those so colonized. A more actively engaged student can
only be engendered via a decolonization process in which power is
circulated more broadly and control is negotiated. But how are most
Ph.D.s to understand this problem, never having learned the critical
importance of teaching before being thrown into the classroom
soon after the dissertation's ink is dry? It was at this point that I
began to recall my high-school days.

This lesson I was learning (again) simply was reinforced by my
experience as a postdoctoral fellow in an institute the goal of which
was to "translate" all the good faculty research being undertaken at
the University of Maine into "terms" that lay people, particularly
politicians, could understand and thereby make some "use" of. This
task, although seemingly off-task for me personally, taught me the
importance of clear, accessible written expression, previously lost on
me in my quest to be an "independent," "original" scholar. If the
message is not heard clearly by other than those already initiated,
what's the point of the production of such knowledge? And this realization, now understood more thoroughly in practice, led to my
realization of what I consider to be the other main obstacle to effective teaching: the dearth of good, accessible textbooks, particularly
at the intermediate level. This dearth is the result of the bias toward
"original" research to be sure—after all, text writing is not consid-

ered research at most institutions—but it also stems from the fact that most of us do not spend enough time working on our writing and determining, more precisely, the nature of the audience we hope to reach. As a result, I think, we now have a plethora of introductory textbooks in geography written at about the ninth grade (U.S.) level, at best, and then any number of "research-oriented" texts that are written in "professional" styles that even my top master's students barely can understand. Who, indeed, are we trying to reach with such messages? (Shades again of Sohn-Rethel 1978.)

Anyway, we spent four years in Maine. I eventually finished my dissertation after many a night and weekend writing, and Ingrid, her Ph.D. degree in plant science, focusing on wetlands. As per our agreement in Leuven, it was now time for me to seek a position based on my specific training with the understanding that the family would follow. Obviously, however, it was not that simple. Ingrid kept her eyes open for appropriate jobs as well and I kept a relatively open mind about where we might end up and why. But we generally was agreed that it was my turn to lead, because of the largely off-task nature of my career at Maine. And so, my first search for a faculty position was on with the usual problems associated with the timing of need as opposed to that of appropriate opportunity. What was unusual was that I felt consciously the need to search for a faculty position as a related individual embedded within a family group with demands and ultimate needs of its own, perhaps even quite distinct from my personal ones. In short, I now experienced more practically the (group) decision-making process Ingrid underwent earlier in search of an adequate Ph.D. program.

And, yet, even with Ingrid's earlier experience, this process proved more difficult for me than I had anticipated. It was not only that it seemed to me to be against my usual and assumed practice as the norm within the social canon. After all, after seven years of practice being related, I was certainly, if still somewhat incompletely, on to the necessity of the group decision. Rather, the problem at the time was that making such a decision seemed still so much contrary to the way I was academically trained that I felt I was somehow selling short my trainers. Being trained as a researcher still seemed to me to necessitate attaining a research-oriented position at a major research-oriented institution, and nothing less. That is what my stu-

dent peers had done and that, I thought in the back of my mind at
least, was what I should do. For this very reason I even applied to,
and interviewed at, one institution where I in no way wanted to be.

Back Again to Tampa, Florida

I suppose this job search is a good indication of how slowly and
unevenly I have learned practically about the implications of marginality. Of the two institutions that actually offered me a position in
my field, USF appeared to be the best fit. But, significantly, I was able
to discern this only after my job search and interviewing process
were well underway. That is, although I thought a research-oriented
institution was the right way for me personally to go, after visiting a
couple of such institutions it was clear that they would not be right
for my family, whether or not I was actually offered a position. Even
at this debut of another century, the academic profession remains
hegemonically biased in favor of single, unattached people who intend to remain such through the tenure process. Very few institutions make any accommodation for partners and attached others,
most often considering such as mere hindrances to the success of
desired candidates. And this bias, of course, is particularly true for academics at the beginning of their careers, who do not command the
big name-power to force institutions to jump to their beat.

It was only after some interviews that I was able better to understand this situation. Each time, I would explain my family needs in
detail and, each time, I would receive the same platitudes about possibilities without any substantial assurances. And, after each time, I
could not help but wonder about the implications of my even having raised the point: was I now a somewhat tainted candidate because of my "personal attachments"? And, indeed, after each time, I
became even more incensed about the structuration of such a bias
against relatedness in academe; a structuration that very few seem to
view as oppressive and even fewer seem to want to change in any
way. Again, USF seemed a good fit, not because the institution actually made any accommodations of this nature for us, but, instead, because it was very young, apparently horizontal and flexible
administratively, and located within a large metropolitan area where

we could possibly find other professional opportunities. Perhaps, we thought, we might even be able to "carve out" a faculty position for Ingrid once there, given the nature of the institution.

To me, and I am sure to some of my closest peers, however, USF seemed an odd choice. True, the small master's program was being completely made over with several new hires and soon-to-be retirements. But the makeover was slated to be oriented toward teaching applied, GIS-flavored geography to large numbers of undergraduates as aspiring planners and consultants. Having been trained in a more liberal-arts fashion, I was destined to a somewhat poor personal fit, and so it has turned out. But after my limited experience on the academic job market, I clearly made the choice for other reasons. Not only did it seem that the institution and place were more prone to allow for more family chances, the very fact that the university and the department were less singularly fixated on research output appeared more likely to allow for family time, even during the otherwise intense tenure-track process—family time, nights and weekends, which I now knew to be critically important for the success of the related group.

And so we came to Tampa, Florida, for all of us the first time we would live in a warm climate and, still, with no assurances of an eventual professional position for Ingrid. After the initial chaos of Ingrid finishing her requirements in June, driving the U-Haul from Bangor to Tampa in late July, finding a suitable residence, and beginning the academic year in mid-August, we began to settle into our new roles. Kieran surely found Florida to his liking, living in shorts and t-shirts all year long in the sun. Initially, as well, Ingrid was content, winding down from her degree accomplishment and now adjunct teaching in the biology department. And I was off on the tenure track, preparing for classes and trying to determine what I needed to publish to make good.

Nevertheless, things eventually became more trying, as Ingrid languished in adjunct-hell, hanging on as I had done at Maine, only more tenuously. She eventually had to take a position outside of academe as a professional within a private environmental consulting firm, which meant long, tedious hours of mostly mind-numbing number crunching. In short, by the third year of my track and after another child, Delaina, it was increasingly clear to us that USF was no more flexible with regard to faculty couples than anywhere else. So, we both began to look for possible faculty positions elsewhere, particularly pressed because the longer Ingrid remained out of the academic loop, the harder we both knew it would be for her to find a suitable position within academe.

Then, just by chance, my best colleague friend at USF, Mark Amen, an international studies professor, became an associate dean in our college. As he was concerned that both Ingrid and I were considering leaving Tampa, he asked me in his new administrative capacity, what would make it possible for us to stay. This question allowed me to articulate more formally our evolving views concerning the plight of faculty couples within the system as it exists. On the basis of on our own experience as well as the previously published experiences of others, I put together a brief report underscoring both the problems that we, as fused professionals, faced, as well as possible solutions USF could pursue in order to alleviate such problems generally, not only for Ingrid and me (see Smart and Smart 1990; Ecological Society of America 1993; Lubchenco and Menge 1993; Wake 1993; Archer 1994). In the report, I argued that if USF were to institute a university-wide formal policy with regard to accommodating faculty couples, the university overnight would be able to attract top-notch faculty candidates it would not otherwise attract; surely a win-win proposition.

Mysticism aside, it was also just by chance that the new dean of our college, David Stamps, originally had come to USF as a faculty spouse and had experienced most of the career indignities Ingrid was now experiencing. Once he received my report he almost immediately notified us that he was going to use us as a case study to determine the worth of the proposition I made with regard to accommodating faculty partners. Fortunately, Ingrid already had made a name for herself with her hard work around campus, so this experiment raised very few hackles when Dean Stamps created, by fiat, a tenure-track job for Ingrid in Women's Studies (which is a whole other story, given her training, but that, again, is for her autobiography to relay). In short, in the space of what seemed like only weeks, Ingrid and I went from searching the country for suitable—connected!—positions to feeling quite satisfied with ourselves for choosing our first faculty position so wisely.

Circles Within Circles

Now I have tenure and Ingrid is in her fourth year of the track. Although I am still a relatively bad fit in my department, I do, I think, play an important role there asking big questions, the answers to which may not be so easily forthcoming or agreed upon. Also, I have become involved in many extra-departmental programs like the one already mentioned, Learning Communities. This latter I helped model after my high-school program (itself modeled after the then newly opened Evergreen State College in Washington State), and it caters now to several cohort groups of about fifty undergraduates who stay together for most of their university credits in the first and second years, collaboratively and interdisciplinarily instructed by five or so faculty from five or so different disciplines. The idea is to create a small, interdisciplinary, liberal arts college atmosphere within our large, metropolitan, essentially commuter institution. One reason for doing this is the recognition that most of those students who drop out of university do so for nonacademic, predominately social reasons. The Learning Community environment is intended to engender a better sense of relatedness amongst students to combat this type of social alienation. My reason for participating stems from the same motive, but also stems from my own trajectory through such a program of classroom decolonization and educational collaboration in high school, a profoundly life-changing experience.

The Learning Community experience thus has reinforced my views of the necessity of decolonizing the classroom, first derived from my practical experience hanging on in Maine. This experience also has brought closer to home the ways in which the classroom can serve as a laboratory of life. When a number of us faculty began the experiment, it soon became clear that our respective understandings of what "community" actually entailed varied widely. The hegemonic understanding was that community generally meant consensus, where "everybody" got along. As a result, many faculty—and participant students—eventually were disappointed with the reality of their Learning Communities as soon as dissensus and actual conflict among individuals began to occur. The similarity between this belief, and its aftermath, and that of new urban "community" planners struck me deeply, allowing me better to recognize many possible avenues that literally connect my classroom activity with my research quite seamlessly. What, for example, is community? How is it formed? What holds it together? What is the role of competition, collaboration, dissensus, and so on, in community development?

In the end, my experiences as a related individual on a fused career path have led me to jobs for which I was not ultimately trained very well to hold and to places where I never thought that I would be. In the process, I have attained much practical knowledge of what I had only grasped previously and unevenly in theory. Indeed, in more quiet times I cannot help but feel that the path taken has turned out to be more circular and ultimately rewarding than I otherwise, quite "normally" would be able to make out.

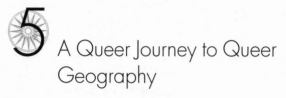

A Queer Journey to Queer Geography

Lawrence Knopp

Antecedents

The seeds of any journey can be found, in part, in formative experiences and processes. These operate at a variety of scales and in a variety of locales and can appear (or feel) highly idiosyncratic or individual. But such experiences and processes may have very broad social origins and/or meanings (Rose 1993; Gibson-Graham 1994; Pile 1996; Women and Geography Study Group 1997). This narrative is intended to improve understanding of the recursive relationships among and between new forms of geographical knowledge and (auto)biographical methods. Hence discussion of these formative experiences and processes, along with more recent ones, is necessary.

I can identify at least four such experiences and processes in my life, each of which has contributed strongly to the development of both my geographical imagination and my theoretical, epistemological, and (activist) political orientations. These are: a sense of myself as *different;* a preoccupation with notions of *justice* and *social change* (including an impulse toward activism); place-based and secularized fantasies of *salvation* and *deliverance* (which included maps as a primary instrument); and a linking of these to idealized (one might even say romanticized) notions of *environment* and *nature.*

Now, the sense of myself as different has to do with various childhood experiences that I took, for one reason or another, to be atypical and that were, to varying degrees, painful. These include, significantly, an early awareness of physical and emotional desires that did not conform to my (learned) expectations. The preoccupation with justice and social change was clearly a reaction—at least in part—to injustices I perceived in this sense of myself as different. But at least as important were other influences and experiences, including the once radical (but later merely liberal), upwardly mobile, post–World War II, masculinist, urban and intensely secular Eastern European Jewish-American experience and identity represented by my father, and the somewhat feminist, middle-class, professional, rural, and educationally oriented white-Christian (though also fairly secular) tradition and identity represented by my mother. The social turmoil and political engagements of the 1960s (strongly represented by my older, draft-aged brothers) were also very important, for they gave me not only a strong (if somewhat abstract) political consciousness at a relatively early age, but also the language and other tools I needed to understand more local-scale and "personal" experiences politically.

Particularly influential were certain experiences associated with educational environments: The first schools I attended were inner-city, multiracial state-funded schools in Seattle, Washington (where I was born and raised, though I also briefly attended a state-funded suburban school in the Netherlands, where my family lived for a year). The social environments at these institutions were, to my unsophisticated child's eye, relatively happy and unproblematic (no doubt they were much more complicated than that, but I did not notice or understand this at the time). However, when I was later placed in a small, privately financed, upper-class and virtually all-white school (located, significantly, adjacent to another multiracial state school where I might otherwise have attended), I suddenly became aware of a quite brutal social hierarchy. As a newcomer—but probably more as a student whose family origins and cultural/political identifications were not typical in the school's elitist (albeit somewhat educationally progressive) milieu—I was positioned at or near the bottom of the hierarchy. The trauma of that experience was quite significant; something that I immediately associated with my peers' obvious class (and to some extent race) consciousness and privilege. Still, I responded contradictorily, by both despising and aspiring to certain of their values, practices, and privileges. This is roughly the same point at which I first recall racism seriously affecting my judgments: While in retrospect I can recall occasional earlier instances of racist thinking and behavior, it was not until this point that they became internalized to such an extent that I actually *feared* the prospect of attending the neighboring state-funded school, almost entirely because of its large African-American student population.

The fantasies of salvation and deliverance, and their linking to maps and idealized notions of environment and nature, were more specific and proactive responses to my early senses of pain and injustice. I sought answers to both social and personal pain in idealized fantasies of families, communities, and other environments. For me, central to these fantasies was what I saw as the cleansing and humbling power of "nature" (and, in particular, dramatic "natural" landscapes to which I saw human beings as most properly being subservient). Maps were the chief tool I used to visualize these fantasies, which were fueled in part by romantic notions about my mother's childhood (which was spent in the rural and small-town intermountain West of the United States) and her related strong "environmentalist" values. Although *both* of my parents supported these values (at least in the abstract), they nevertheless stood in stark contrast to my father's working-class Jewish origins in the urban South (from which they had both fled quite early in their lives together), as well as to many (but by no means all) aspects of the urban world of Seattle where I lived. Indeed, the allure of these fantasies was so strong (especially in light of ongoing personal pain) that by the time I was an early teenager I had become fixated on recovering for myself what I imagined to be my mother's idyllic past. I obtained and studied topographical maps of the places about which I fantasized, and arranged to spend (with my parents' grudging acquiescence) a year in my mother's small home town, living with a local family with whom she remained distantly acquainted and attending the local state-funded secondary school.

That experience, although generally quite positive, began a long process of demystifying and deromanticizing my notions of "nature" and nonhuman physical environments as redemptive forces. The family and community in which I lived, while in many ways conforming positively to my hopes and expectations, were nonetheless dominated by more abstract political and social values that were, in the main, diametrically opposed to my own. And, of course, the experience in no way offered answers to the dilemma represented by my increasingly urgent homosexual desire. I began then to learn that it was not only my own experience that was characterized by contradictions, but also that pain and injustice could not be swept away simply by altering environments.

Stumbling Toward Geography

The lesson took years to sink in, however (and may never fully do so). After returning to Seattle, I successfully carved a niche for myself as an intellectual of sorts who was a member (and at times a leader) of a mildly countercultural (but mostly, in retrospect, liberal) social faction at the privately financed, secular, and educationally progressive college-preparatory secondary school that I now attended. As an unofficial in-house cultural and political critic, I enjoyed the support of family and even school administration, all of whom were affected not only by an educational philosophy that encouraged critical thinking, but by the larger social and political events of the time (the early 1970s).

But the pain and injustice both of my closeted homosexual desire and of the larger social world did not abate. When it came time to think about college (the idea of doing anything else did not even seem to be an option), I again sought answers in a change of environment. I selected a less prestigious Midwestern liberal arts college over more prestigious and "environmentally" agreeable (but also more conservative) ones on the East Coast, in the secret hope that it would prove a place where I could at last come out as a gay man. It (and I) did not, so after two years (including a term on a language program in Europe—another failed attempt to find answers in a change of environment), I left college altogether and returned to Seattle and, more importantly, the more environmentalist strategies of my early teenage years. I began spending time and doing odd jobs on an isolated island I had visited many times in northern Puget Sound, where family friends had vacation and retirement cabins and homes. Eventually I moved there full-time, living spartanly, with neither power nor plumbing, in a one-room cabin that I rented for next-to-nothing from absentee owners. I became peripherally involved in the local community of about two hundred people, solidified what was already a deep attachment to what I perceived as a nearly idyllic "natural" as well as social environment, and contemplated making this life as a countercultural urban expatriate permanent.

But the failure of even this near-idyllic environment eventually revealed itself. The local community, although pleasant and largely united in terms of environmentalist values (which I shared), was stratified, exclusivist, and socially limited (and limiting), offering nothing of consequence in the way of support to me as a potentially "out" homosexual. And compared to my always challenging educational environments, it was quite unstimulating intellectually. So with some sadness (for this was the closest I had ever come to realizing my fantasy) I again returned to Seattle, this time sharing housing with my newly divorced older brother and finding work in the burgeoning restaurant business of the mid-to-late 1970s.

While this new environment also offered less than I was used to intellectually, it was (at first) socially quite exciting. The restaurant business was filled with large numbers of young openly gay people as well as others living nontraditional or transient lifestyles. My brother's young professional, urban, heterosexual social world was also quite stimulating (though I eventually moved into an apartment of my own). And, when almost immediately the former beauty queen and celebrity Anita Bryant's nationwide antigay "Save Our Children" campaign came to Seattle, in the form of an attempt to repeal a recently passed nondiscrimination ordinance that included protections for gays and lesbians, my opportunity to come out finally presented itself. I prepared those around me by becoming active in the effort to defeat the repeal, which seemed to them a perfectly ordinary act, since I had been interested and involved in electoral politics virtually my entire life. So in that context, at twenty-one years old, I finally came out to family and friends and began exploring gay relationships and what I viewed as the gay community. The attempt to repeal the gay-rights ordinance was defeated by a two-to-one margin, and the ensuing collective as well as

personal euphoria led me this time to romanticize about the redemptive potential of certain aspects of *urban* environments.

But gay life and my new urban gay environment ultimately also
failed to satisfy. The social and cultural milieu of both the restaurant
business and the gay community eventually offered little new to me
as a gay man, and even less in the way of a future. And both seemed
singularly apathetic about wider social injustices. I experimented
with mid-level (and seemingly gay-friendly) service-sector office
jobs, in the hopes that both opportunities and consciousnesses
would be more developed there. But the reality of these jobs, like the
restaurant work, was also disillusioning, unstimulating, and limiting.
All of this led to a sense that my personal growth generally, and my
coming-out process in particular, were somehow still incomplete.

So after a four-year hiatus I returned to school, this time at the
University of Washington in Seattle, as a working, self-financing,
nontraditional, commuter student. Knowing nothing about geography as a discipline, I chose to continue studies in political science,
largely because I had accumulated more credits in that field than any
other and because it seemed to make the most sense, given my lifelong preoccupation with, and involvement in, politics and social justice issues. But I had a very pragmatic motivation, too, in that I was
determined to develop marketable skills that would free me from
what now seemed to be not only dead-end career opportunities, but
limited personal and political ones as well.

Once again, then, I returned to my interest in places and environments, this time from a more applied, academic perspective. In particular, I turned to maps, which I had always loved contemplating
but now also saw as extremely practical tools. I sought to develop
skills that would allow me to use maps in the realm of politics, thus
making me marketable as a political strategist, consultant, or (ide-
ally) activist. I studied cartography, computer cartography, and spatial analysis (under the auspices of the geography department—my
first exposure to the discipline!) along with required courses in political science. I also continued to involve myself in mainstream electoral politics, showcasing my developing talents in spatial analysis
along the way (in the naïve hope that I might be able to market
them in such a way as to maintain my personal and political in-

tegrity as well as freedom). What I found, however, was that the discipline of Political Science (in that time and place, anyway) had little use for someone interested in either social justice or spatial analysis (much less a combination of the two), while the core culture and values of mainstream politics (including so-called liberal politics) were increasingly revealed to me, at the dawn of the Reagan era, as cynical, fraudulent, and repugnant. I was, without fully appreciating it, becoming radicalized.

Near the end of my undergraduate career, a classmate and our mutual instructor in a spatial analysis course encouraged me to consider graduate school in geography. On their advice, and still knowing relatively little about the discipline of geography, I applied to two programs (one local and one nonlocal). My goal was still to become an applied (activist) spatial analyst. By then, however, I knew that I wanted to apply my skills in a variety of activist forums, not just (or even primarily) in mainstream electoral politics (I had little, if any, idea what these might be, however). To my surprise, I was accepted into, and offered financial support at, both programs—in spite, significantly, of my having found a way to explicitly signal the facts of both my activism and my homosexuality in my applications.

The prospect of graduate school seemed an exciting alternative to my now rather stagnant and otherwise apparently limited future in Seattle. So I opted not only to go, but to accept the nonlocal offer. And although my decision was again partly (and most likely subconsciously) grounded in a hope that new answers would be found in a new environment, the environment to which I was relocating (Iowa) was one that I considered—given the way I had always conceived of and prioritized "natural" elements of environments—anything but ideal. The hope now (as it was when I left for college the first time) was that a change of environment *in itself* would prove to be enlightening and liberating, and oddly enough I looked for this in the aesthetically unremarkable (compared to my home region) and relatively conservative social milieu of the U.S. Midwest. My lifelong conception of "nature" and "natural" landscapes as external, redemptive forces in human social and personal life was now at its lowest ebb. Instead, I looked more broadly to *difference* as the most salient redemptive feature of environments.

A Space for Sexuality

Graduate school was an enormously broadening and liberating experience. I began reading about the history and philosophy of the discipline of geography, and it began to dawn on me that there might in fact be broader opportunities and applications available in the field for someone like me than I had realized. Particularly influential, initially, were readings in Marxist and, to a lesser extent, feminist social theory as these were being interpreted and applied by geographers. These included seminal works on urban social theory and urban political economy; on the capitalist state; marxist, feminist, and emerging poststructuralist perspectives on social reproduction; and various philosophical and methodological critiques of positivism in social science.[1] This conjuncture of critical ideas, grounded in critiques of capitalism, positivism, and, to a lesser extent, patriarchy, provided me with the language I needed to explore more seriously my long-standing interest in questions of social justice as they pertain to gender, sexuality, space, and place. Conspicuously absent from this base, however, was any body of literature theorizing "nature" and its relationship to these issues. Also conspicuously absent was a central concern with spatial analysis of the more technical variety, for which I had, primarily, selected my graduate program. So these concerns were put, as it were, on hold.

By this time (the mid-1980s), there had been only a handful of attempts in geography to address gay and lesbian issues. These included perhaps most significantly the groundbreaking, though

---

1. Harvey 1973, 1985; Edel 1981; Saunders 1981; Castells 1983; and Paris 1983 influenced me a great deal with regard to social theory and urban political economy. Work on the capitalist state include Giddens 1979, 1981; Clark 1981; Dear 1981; Dear and Scott 1981; Fincher 1981, 1984; Reynolds 1981; Duncan and Goodwin 1982a, 1982b; Fincher and Ruddick 1983; Clark and Dear 1984; and Lauria 1986. Impelling readings in radical literature in marxism, feminism, and poststructuralism comprised Zaretsky 1976; Holmstrom 1981; Castells 1983; Mackenzie and Rose 1983; Buechler 1984; and Rose 1984. Finally, the works influencing my ideas within the radical literature on the critiques on positivist science are Keat and Urry 1982; Sayer 1982, 1985; Fincher 1983; Thrift 1983; Chouinard, Fincher and Webber 1984; and Gregory and Urry 1985.

largely ignored, contributions of Robert McNee (1984, 1985), William Ketteringham (1979, 1983), and Barbara Weightman (1981). Of these, only McNee's contributions were informed by any kind of radical social theory (in his case, anarchism). And again, these had not been taken seriously by any community of academic geographers with power, at that time nor several years later. So it became clear to me that an opportunity *might* exist to pursue my conjoined personal *and* academic interests in the context of the strong emerging concern with questions of social reproduction, gender, and culture among critical urban geographers. I focused in particular on the empirical context of gentrification, since this was a flashpoint field at the time for such questions and concerns. It also quite clearly impinged on questions of specifically gay cultural identity and expression, which gave me the opportunity to broaden my reading in what at the time was known as "gay and lesbian studies."

I read as much literature on gentrification as possible, and also read widely in feminist and lesbian/gay social theory.[2] In the spring of 1984 (barely a year into graduate studies), I presented at a departmental colloquium a theoretical model of the relationship between gay community development in cities and gentrification. It was a terrifying moment. Obviously, the act constituted a significant step in my coming-out process. But it was more than that. Not only was I signaling to all those with any doubt that I was gay, I was signaling my intention of making this fact central to my own intellectual project for the foreseeable future, which of course meant that the department would have to grapple with it as well. In all likelihood, the decision to make the presentation—in which, as I recall, no one except a very few graduate student peers was consulted—would either make or break my graduate-school career.

To my considerable pleasure (and relief), the presentation was well received. Indeed, as I recall its success was a major contributor to my decision (encouraged by certain faculty and graduate student

2. Works on gentrification included: Smith and LeFaivre 1984; Smith 1979, 1983; and Rose 1984. In lesbian/gay theory, crucial works were Levine 1979; Foucault 1980; D'Emilio 1981, 1983; Altman 1982; and Snitow, Stansell, and Thompson 1983.

peers) to change status from M.A. to Ph.D. student. And ultimately it led to collaboration with an early and important mentor in the form of a published work (Lauria and Knopp 1985). From that moment on the rest of my graduate career was defined: I would write a dissertation in which a refined version of this model was applied to at least one (and possibly more) gentrified or gentrifying neighborhood(s) in a U.S. city.

With the help of two mentors and many other supporters, I did so. In the process, I temporarily changed environments, yet again, to inner-city New Orleans, Louisiana, where I lived for the better part of a year conducting field work. Although professionally I salvaged "success" from the experience, it was in fact much more mixed, both professionally and personally. Privately I saw it as a fairly transparent attempt to further work out my own (gay) identity while at the same time pursuing an academic agenda. I still did not realize how inseparable—and, indeed, nearly identical—the two were. And although I learned much of value there, some of the most important lessons about space, culture, and sexual identity (regarding, for example, urban anomie) never made it into my dissertation. Even so, the work was well received within the broader disciplinary circles to whom it was largely addressed. As a result, it became increasingly clear to me personally that a career in academia (and an opportunity to continue working out issues of identity, culture, space, and place personally as well as professionally) was a possibility. So long as I continued, in the main, to enjoy my work, this seemed an attractive goal.

As I was finishing my Ph.D., I was fortunate to receive a tenure-track position in the Department of Geography at the University of Minnesota-Duluth. The position was for an urban geographer, but its appeal was as much due to its location in a relatively peripheral, low-population-density location, with dramatic scenery and a difficult-to-modify physical environment, as anything. At some level, I was back to imagining (or hoping for) an idyllic fusing of physical environment and human community. And the idea that I could continue pursuing my interests in urban gay culture and politics, while immersed daily in a more mixed environment with opportunities for local involvement, did seem almost ideal.

The first few years of the job, however, were focused almost entirely on *securing* the position (by further establishing my niche in the discipline and, ultimately, achieving tenure). Ironically, this meant spending very little time engaging or establishing ties in the local community. Instead, I wrote, taught, and pursued new international (rather than local) research opportunities. When I was fortunate enough to win a career development award that would take me to the United Kingdom and Australia for a year, I jumped at it. So it was not until I returned from the year abroad—nearly five years into the job—that I really began to engage seriously with my local community and environment.

First, however, there was the experience of a year abroad. In the United Kingdom, I was challenged intellectually and personally to think beyond the rather mechanistic and United States–centric models that had until then informed my work. It was a time when postmodern, poststructuralist, and postcolonial theories were rapidly coming to the fore in feminist and lesbian/gay studies and reinvigorating cultural and political geography as well (e.g., Harvey 1989; Jackson 1989; Soja 1989). In addition, the field of lesbian and gay studies was being challenged by its own postmodern variant, queer studies (see below). I listened, read, learned, wrote, and spoke a great deal over the course of the year, while exploring queer and mainstream environments in settings as diverse as urban and rural Scotland and England, Finland, the Netherlands, and Australia.

The result was that writers now spoke to my own experience and shaped my thinking, at least as much as people like David Harvey and Neil Smith.[3] Accordingly, my written work took a significant turn, not so much away from urban political economy as *toward* urban *cultural* political economy (see, e.g., Knopp 1994, 1997, 1998). I began linking gay cultural and political spatial forms, for example, to geohistorical projects such as nation building. What was still missing, however (and remains absent, for the moment), is a well-developed theory of the relationships between these (and, indeed, all cultural forms) and *physical* environments. That project still awaits.

3. These writers include: Butler 1990; Enloe 1990; Sedgwick 1990; Young 1990; Wilson 1991; Bondi 1992; Davis 1992; Grosz 1992; Gilroy 1993; Rose 1993; Bell et al. 1994; Chauncey 1994; Kobayashi and Peake 1994; and Lake 1994.

Toward a Queer(ed) Geography

In the meantime, I have been reading "queer" cultural studies of various kinds, including nonacademic work such as Browning's *A Queer Geography* (1996) and Fellows's *Farm Boys* (1996). Much of this work is informed by a geographical sensibility and imagination, as well as by academic disciplines such as sociology, psychology, and literary criticism, but, interestingly, *not* by academic geography. Although at first glance this seemed strange to me, I have since come to see it as entirely understandable. My analysis (inspired by certain poststructuralist geographers, e.g., J.-K. Gibson-Graham [1999]) is that many critical geographies (like critical approaches to international politics—see Enloe 1990) have unintentionally tended to reproduce many of the structures of power that they supposedly seek to subvert (as well as others, such as heterosexism, ageism, and ableism, which they only address at best tangentially).

The "political economy of urban space" approach is a case in point.[4] Its importance to my own thinking and work cannot be overstated, and I continue to be amazed by its tremendous flexibility and adaptability (in terms of accounting for the proliferation of "difference" in the contemporary world), which only adds to its formidable analytical power. Still, I am persuaded that this tradition conceives of socially relevant forms of human difference as arising almost entirely from, and being virtually completely subservient to, a dynamic derived exclusively from class relations, which in turn are conceived in disembodied form (whereas, ironically, the abstraction "capitalism" is itself re-embodied—or at least gender coded—by the use of language such as "colonization" and "penetration" [again, see Gibson-Graham 1999 on this point]). As such, the tradition can hardly by termed feminist, much less "queer."

Many "mainstream" feminist alternatives, meanwhile, strike me as exemplifying a somewhat different sort of unintentional conservatism.[5] Whereas I continue to be awe-struck by the broad applica-

4. This approach is exemplified by Harvey 1973, 1985, 1989, 1990; and Smith 1979, 1983.

5. These works include, for example, McDowell 1983; Mackenzie and Rose 1983; Lewis 1984; Massey 1991b.

bility and analytical power of their focus on the importance of understanding class and gender relationships' mutual and historically/geographically contingent constitutions, I have come to see that they can also be quite heteronormative and, at times, disempowering (even to feminists).This is due to the fact that they tend to conceive of "gender relations" as male-female relations (even though the social constructions of masculinity and femininity are as much about regulating and controlling the relationships of men with one another and women with one another as they are about regulating men's and women's relations to one another) and tend to construct capitalist patriarchy (or patriarchal capitalism) as a unitary and utterly hegemonic social formation.The trouble with the latter is that such a representation, which ignores the much more diverse lived experience of gender and class relations, reproduces and normalizes capitalist patriarchy's hegemony.

I find myself in a bit of a dilemma, however, since I also have troubles with some of the more postmodern, queer-theory-influenced alternatives to these more mainstream traditions (e.g., Gibson-Graham 1999; Bell et al. 1994; Kobayashi and Peake 1994). Some of these, it seems to me, tend to regard *any* kind of abstraction, *any* kind of categorization, and *any* kind of generalization as ill advised and politically conservative. In the name of honoring the polyvocality, fluidity, and fragmentation of "real" life, David Bell, Jon Binnie, Julia Cream, and Gill Valentine (1994), for example, refuse to draw any conclusions, or to take any significant political positions on matters as highly charged as the racialized sadomasochistic performances of some gay male skinheads. Similarly, some other geographers (myself included, at times) have perhaps become so preoccupied with celebrating the "queer" accomplishments of some nonwhite, working-class or female gentrifiers that they appear almost to dismiss "old-fashioned" concerns about capital accumulation and class exploitation in the gentrification process as buying into a racist, sexist, and heterosexist (not to mention "totalizing") discourse.

But to me this can amount to a sort of postmodern "puritanism," which is at best unnecessary and contradictory; at worst, an abdication.The problem is that a legitimate concern with the political implications of certain modes of analysis and methods of discourse has led to the viewing of virtually all politics, *except* the politics of deconstruction, as suspect. Not only is this a contradiction (in that the politics of deconstruction are themselves grounded in a set of larger political principles; see Fuss 1989), it also leads to a situation in which the only remaining social value is raw power. And that bothers me very much. Too often in this sort of approach one set of pressing political issues (for example, the links between race, class, gender, and sexuality) end up being ignored because of a fear of selling out on the discursive front (i.e., not adequately honoring fluidity and fragmentation, engaging in the essentializing of differences, or buying into a totalizing discourse of some kind or other). Now, I do not mean to suggest that discursive issues are less pressing or practical than more "material" ones (indeed, as should be clear by now, I am inclined to reject this distinction altogether). Rather, I have come to prefer engaging on both fronts in a principled and pragmatic way. This means avoiding both discursive *and* materialist "puritanism." Otherwise I fear that those who crave power for its own sake (or for the sake of principles we know we oppose) will not hesitate to fill the vacuum.

"Activating" Queer Geographies

My most recent concern, then, has to do with defining a "queer" critical geography that has something substantive to offer politically. How can such a critical geography be translated into meaningful political action "on the ground," either within or beyond academia? I have discussed elsewhere how this might be accomplished in the context of "research" (Knopp 1996). My argument there has basically been that research questions themselves must be formulated in ways that are self-consciously political, geographically and historically contingent, and that conceive of social "reality" as ever-changing, always incomplete and characterized by cross-cutting, mutually constitutive, and frequently contradictory sets of social practices and relationships. "Research," then, becomes a *critical, interpretive,* and *contextualizing* enterprise as much as a traditionally "empirical" one. In the context of teaching, service, and "activism" more broadly (both within and beyond academia), the same principles apply.

In the communities of which I am a part in Duluth and at the

University of Minnesota-Duluth (UMD), for example, coalitions made up of students, faculty, staff, and people from outside the university have been involved for several years now in a number of critical interventions in the ways that various actors in the area (formal, informal, institutional, organizational, and individual) conduct affairs. The goals of making local cultures and practices more democratic and inclusive have always been explicit in this activism, and the particular battles and strategies associated with these have always been tailored to local contingencies and political "realities." Here are a few examples.

When a hate-speech-filled "parody" issue of the free campus student newspaper was published in 1996, an informal and diverse coalition of African-American, lesbian, and other student activists (with support from some faculty, staff, and nonuniversity community members) collected the vast majority of the issues from distribution points, removed the parody insert and redistributed the offending material at a protest rally the next day. This set in motion a chain of events that briefly commanded national attention and highlighted the particular ways in which the contemporary American "culture wars"—in all their contradictions—are manifest in the rather offbeat, peripheral, and quirky social spaces that are UMD and Duluth. It also highlighted the highly contingent and fragile set of political opportunities and constraints that exist in this particular locale: The university administration formally supported the protesters, while the university community as a whole, along with most of the nonuniversity community, appeared deeply divided by a tug-of-war between the equally strong, deeply ingrained impulses toward forms of radical individualism (manifest in an "anti–political correctness" crusade) and local forms of collectivism and communitarianism (known locally as "Minnesota Nice"—see the Coen brothers' brilliant film *Fargo* for a hilarious depiction of this). The result has been increased formal opportunities for minority-cultural expression in the paper (as well as in other university and nonuniversity institutions), at the same time as there has been continued resistance to virtually any and all substantive changes in *practices* at the paper and elsewhere demanded by activists (e.g., implementing forms of affirmative action in the hiring of student editors and changing the name of the paper from "Statesman" to something gender-neutral).

Similarly, UMD's small but exemplary academic program in women's studies has established itself as an important nexus for political activism within and beyond the academic community. It enjoys a "baseline" level of at least minimal support from university administration (albeit less, in a material sense, than most other programs) and uses this position to support a wide range of feminist and progressive initiatives. Like all women's studies programs, it is periodically vilified by opponents who brand it as "ideological," "narrow," "antimale" and "dominated by lesbians." This is most likely to happen when students, faculty, or allies engage in critiques of practices *outside* of women's studies. When I initially proposed a course in "Feminist Geographies" to my college's curriculum committee, for example, I was greeted with arguments to the effect that such a course was "too specialized" for an undergraduate program, that there were already "too many" courses with feminist perspectives in the university's curriculum, that women's studies was not a program to which geography (or any other department) would be wise to build connections and that *obvious* connections between feminism and geography did not exist! Similarly, women's studies students and faculty are consistently among the most outspoken critics of university and community initiatives in the area of "diversity." We frequently challenge, for example, definitions of and approaches to "diversity" that do not acknowledge the existence of power relations. Although we have thus far failed to make power relations central to institutional approaches to "diversity," we are generally successful in our efforts to get courses with feminist content into the curriculum. After some considerable debate, for example, the "Feminist Geographies" course was approved, as was another on the "History of Black Resistances in the Americas" (that course has never been taught, however, owing to the resignation of the instructor who proposed it).

In terms of specifically "queer" political activism, there are a number of forms of activism in which I, as a geographer, have been involved at UMD and in Duluth. One is the conducting of an annual "Tour of the Lesbian, Gay, Bisexual, and Transgender Twin

Ports" ("Twin Ports" is the local appellation for the Duluth, Minnesota/Superior, Wisconsin, metropolitan area) during Geography Awareness Week. This event is advertised both on and off campus and involves visits to local queer-owned and -operated or queer-friendly sites and venues including a gay men's community center, a lesbian community center, a women's health center, a lesbian-owned, women's-oriented coffee house, a queer youth group, a gay-owned B&B, three bars, private homes, and neighborhoods. During the tour we discuss the need for safe spaces and spaces where queer expressions of desire and cultural identity are accepted, different ways in which queer people construct and use space, the problematic nature of the "public-private" distinction, differences between gay male, lesbian, bisexual, and transgender cultures and identities, and related issues. Thus far the event has drawn little explicit criticism, though there may be a perception on the part of some homophobic students, staff, faculty, or nonuniversity community members that it is part of strategy on my part to "advance a gay political agenda." (I was actually told once that such a perception exists!)

Another form of activism in which I, as a critical geographer, have been involved has been through my service on departmental, collegiate, and university-wide committees. At the departmental level, I have advocated strongly the diversification of our curriculum and the transformation of certain aspects of the culture of our department. This has meant many things, including advocating (and offering!) courses in "Feminist Geographies" and "Geography and Social Justice," incorporating "queer" subject matter and perspectives into my own courses in "Urban Geography," "Political Geography," "Economic Geography," "World Regional Geography," "Geography of United States and Canada," "Urban and Regional Planning," and "Geographic Thought," advocating the inclusion of such subject matter and perspectives into colleagues' courses, arguing for language identifying "culturally diverse perspectives" as "desirable" (or even "required") in candidates for faculty positions, and critiquing certain more nebulous, informal departmental practices (such as our interpersonal interactions, especially with students) that may contribute to a culture that is perceived by women and minorities as less than welcoming. Such advocacy and critiques are controversial, to be sure, but again the dominant liberal political culture of the university has meant that it has been very difficult for a well-argued case on my part to be ignored.

My activism (as well as scholarship) at the departmental level (along with my relatively new status as department head) has led to a certain amount of visibility, within both the university and the local Duluth communities. This, in turn, has led to my appointment to various collegiate and university-wide bodies, as well as to a certain status within the Duluth community as one of the people to whom the local media and others come for comment and input on issues related to the queer communities and "diversity" issues generally. Following the controversy surrounding the parody issue of the student newspaper, for instance, I was appointed to the university's Board of Student Publications, where I am most definitely a thorn in the side of the newspaper staff! In addition, four other queer department heads, numerous student and staff activists, nonuniversity queer activists, and I have critically addressed the university administration and bureaucracy, both formally and informally, on queer issues in contexts ranging from freshman orientation (and student life generally) through personnel and employment practices to campus planning. We speak regularly to teacher-education and other classes (at UMD and even, occasionally, in local high schools) about queer lives and experiences. We have organized forums on issues related to gender and sexuality in the military and within religious communities. And we hold highly visible annual Pride, Coming-Out, and Valentine's festivals, both on and off campus. Although little changes in terms of institutional practices, we have managed to gain several symbolic concessions from institutional representatives of the dominant culture both in the university and in the broader community. We march with little or no incident, for example, in both the corporate-sponsored annual St. Patrick's Day and "Christmas City of the North" parades through downtown Duluth and a recently reopened and politically very "out" gay bar was acknowledged recently with a public ribbon-cutting event by more than a dozen representatives of the Superior, Wisconsin, Chamber of Commerce!

All of this reflects the peculiar local conjuncture of institutional bureaucracies that are quite hegemonically liberal, reactionary as well as progressive populist traditions of radical individualism, religious communities involved in internecine battles between doctrinal liberals and fundamentalists, various ethnically based forms of social-democratic collectivism, and relative racial homogeneity. These combine locally to produce significant opportunities for the development of certain minority-culture political and "lifestyle" communities, their partial integration into public affairs, and even their capacity to influence and change (in a limited way) elements of the dominant culture(s). Such opportunities are, of course, unequally distributed. Individuals such as I, who enjoy race, class, gender, occupational, and status privilege, are obviously in the strongest position to benefit from them. But I would argue that to a considerable extent the savvy exploitation of these opportunities, by various kinds of activists, has resulted in significant improvements in the quality of life for several (but by no means all) minority-culture communities in the Duluth area over the past several years (especially the lesbian/gay and feminist communities). In many ways, local activists have been *doing* for years what critical geographers informed by a queer sensibility and theoretical orientation, such as I, are only now figuring out and advocating: That activism, to be meaningful and at all effective, must not only be sensitive to and knowledgeable about local contingencies and realities, it must also be comfortable living with contradictions and a sense of incompleteness.

What's Yet to Be "Activated"

The preceding should not be construed as suggesting that a queer critical geographical perspective is only a johnny-come-lately to the world of pragmatic activism or that it has nothing substantive to offer nonacademic activists. On the contrary, I see the relationship between academically and nonacademically based activists as mutually informative and beneficial. As Audrey Kobayashi (1994, 78–79) argues:

> [I] am personally committed to acknowledging my research as political and to using it most effectively for social change. . . . [But] the question of "Who speaks for whom?" cannot be answered upon the slippery slope of what personal attributes—what color, what gender,

what sexuality—legitimize our existence, but on the basis of our history of involvement, and on the basis of understanding how difference is constructed and used as a political tool.

There is nothing about the subject position of either "academic" or "nonacademic" that makes it impossible for one or the other to develop such histories or approach such understandings. In fact, it is entirely consistent with a queer conceptualization of social reality to conclude that *multiple* understandings, drawn from multiple subject positions, can be politically valuable. With this in mind, I conclude with a few comments about how a queered critical geographical perspective might further inform local-scale activism in the Duluth area. The implication, of course, is that while Duluth is unique, the mutually reinforcing relationship between academically and nonacademically based activists is something that might, in some general senses, be applicable to other empirical contexts as well.

On matters of race, UMD and the Duluth area are, frankly, much less progressive than on matters of gender, class, and sexuality, or even ability/disability. The demographics of the area overall, and of the local activist communities, are overwhelmingly white-European. The largest nonwhite minority is Native Americans, who constitute less than 5 percent of the local population. Other people of color *together* constitute barely 1 percent. Services for people of color are virtually nonexistent, racist practices such as housing discrimination are rampant and efforts to enact local legislation outlawing such practices consistently fail. Nonwhite in-migrants frequently report an extremely chilly reception from the "locals."

A queer geographical perspective, as articulated academically, might well have something to offer local activist projects when it comes to fighting these forms of racism. In particular, a similar praxis of *critique* and *demonstration,* including many of the same strategies that have been so effectively used to advance feminist and lesbian/gay/bisexual interests locally, might be deployed in the service of antiracist politics. The adoption of a critical and interventionist posture with respect to the *internal* workings of the predominantly white activist communities, for example, would likely result in the opening up of certain opportunities (albeit limited ones) for nonwhite activist community cultures to find a voice within those activist circles. It may not result in certain other desirable changes (especially in the *practices* of these activist communities), but that may simply reflect a limit, for now, of what is possible locally (just as local institutions have not responded, in the main, to lesbian/gay/bisexual/transgender interventions by altering many of *their* practices). And it may well bring about subtle changes that lead to a social climate in the future where people of color grow as a proportion of the total population and feel empowered to press issues in a way that dominant cultures simply cannot ignore.

By queering our understandings of "race" in the abstract, then, in a way similar to, yet more narrowly academic than, the ways in which we queer understandings of gender and sexuality, it may be that an environment can be created that, in the long run, serves the goal of undoing racism. It's a never-ending project, to be sure, and the critical lens of queer geography will ultimately need to be turned inward onto itself. But if a locale as historically homophobic and heterosexist as Duluth, Minnesota, can end up with queer floats in its family-oriented downtown Christmas parade, Chamber of Commerce–initiated ribbon cuttings at gay bars, and more openly queer department heads at the local 7,800-student university than at the 50,000-student university campus in the Twin Cities, are Latino, African-American, Native-American, and Asian counterparts so unthinkable?

"You Want to Be Careful You Don't End Up Like Ian. He's All Over the Place"

Autobiography in/of an Expanded Field

Ian Cook

Situate Yourself . . .

OK. I'm writing this chapter at a strange time. It's March 1998. I'm a geography lecturer at the University of Wales, Lampeter. It's seven and a half years after starting my Ph.D. research, three and a half years after first submitting my Ph.D. thesis, and I'm waiting for the official confirmation from Bristol University that I can call myself "Dr. Cook." Just before Christmas last year, I received a letter from the Higher Degrees office saying that I would be awarded my Ph.D. if I corrected some "errors of substance" in my resubmission. I did this in January and sent a hard-bound copy to the university as requested. So, at the moment, I am "Dr. Cook" on the sign on my office door, on my department's web site, on my most recent checkbook, and on the CV that I submitted with my latest job application. I'm not sure how fraudulent this is. I haven't dared to ask. It's a strange time.

Thanks go to all co-habitants of this expanded field for giving me so much to think about, to the ESRC and Lampeter's Geography Department for funding many of the travels through it, and to Pamela Moss and Jon May for helping all of this to travel further afield.

My thesis had, in large part, been a critique of the politics of academic knowledge, which had made it very hard for me to do a Ph.D. But, I'm currently on the verge (maybe, hopefully) of being awarded a Ph.D. for writing about this. Yes, my "antiestablishment" thesis is on the verge of being given the stamp of approval by part of the "establishment" that it criticized. So, once I can "really" call myself "Dr. Cook," I can no longer be a graduate student railing against "the system." Instead, I'll be part of it: stamped, approved, and able to trade in my plain black "mortarboard" for a rather fetching velvet hat (maybe, hopefully) for the next graduation-day procession at Lampeter. Even if worn at the jauntiest of angles, this hat won't hide the fact that I'll be well and truly part of "the system" by then. And, of course, it's not just a matter of hats. Part of the critique in my thesis had concerned the politics of Ph.D. supervision. During the *viva* (oral examination), my examiners had argued that supervision was often a difficult experience for supervisors as well as for their students. They told me that I'd understand this better when I had my own students to supervise. I didn't then, and I didn't by the time I resubmitted the thesis in May of last year. However, as I write this almost a year later, I'm six months into supervising "my" first two Ph.D. students. And someone has asked me to write a chapter about my Ph.D. for a book on autobiography in geography. This is a time of transition.

What's All That Got to Do
with Autobiography, Though. . . ?

Oh. That's a hard question to answer. And I don't want to answer it just yet. You need to know what my Ph.D. research was about, first. To cut a long story short, I intended to trace connections between the retailing of one kind of fresh tropical fruit that was being sold by the major British supermarket chains in the early 1990s and the people who were involved in growing it on two Jamaican farms. This was an ethnographic project in which I wanted to make, and think through, connections between the overdeveloped and underdeveloped worlds, between rich and poor, between production and consumption, and between the everyday lives of people working throughout a commodity system (see Cook 1995). In the process of doing this research, though, I passed through and was connected with all kinds of other locales (physically and otherwise). And all of these movements and connections had affected what I had been able to do, to learn, and to write.

There was, for instance, my departmental "locale" where, among other things, I had been under pressure form the very start to submit my Ph.D. on time, where I was struggling to convince my supervisors that I could do this with the research I planned to do, where I had to justify my study within often unfamiliar academic literatures, where I also became convinced that I was failing to perform cleverly enough in front of them and their colleagues, and where these kinds of tensions were continually under discussion among the graduate student community, within and beyond the department. There was the locale occupied by the funding body that paid my fees, gave me a maintenance grant, provided my research expenses, and set the outer bounds of my research through regulations that, for instance, allowed funding for just one overseas research trip and one overseas conference. On another scale, this body (the U.K.'s Economic and Social Research Council or ESRC) also largely determined and determines the structure of graduate human geography research in the United Kingdom through setting performance targets for departments concerning, amongst other things, completion times and completion rates for Ph.D.s. ESRC's ability to impose sanctions on those departments that didn't and don't meet these targets explains why most if not all fulltime human geography graduate students in the United Kingdom were and are faced with the pressure—whatever the topic of their research—to submit a thesis within four years of starting (a.k.a. "on time").

So, when it came to doing the preliminary research with fruit-trade people working in the United Kingdom, these locales became folded into each other in intricate ways. For instance, I always had plenty to report when I returned to Bristol after talking to the contacts in the fruit trade that I'd taken time to develop. But my supervisors never seemed to be as excited about what I had found out as I was. What these fruit people had told me was all very interesting, but would it produce a Ph.D. that would get done on time? What *exactly* was I going to study and how was I going to do this? Time was slipping by, I still hadn't told them and wouldn't tell them. But I

wasn't prepared to produce the detailed research plans they demanded—like timetables for doing the research, or chapter outlines
for the thesis—when I knew so little about how British supermar-
kets sourced their fresh fruit. Without knowing much more, I
couldn't even begin to plan out my research. And then there was the
issue of who would let me talk to them about which parts of this
business, anyway. I didn't know what was practically possible. So, I
couldn't even come up with a simple research question. This was a
problem, then (and so was I). I had had to register for an M.Phil. and,
if I didn't get my research sorted out by the end of my first year, I
would not be upgraded to a Ph.D. and would not be funded to continue my study. The completion times and rates of M.Phil. theses
were not counted by the ESRC in its recognition exercise. Not up-
grading an M.Phil. was a "way out" for all concerned. And, that year,
my supervisors reminded me of this alternative almost every time
we met to discuss my research.

These continual reminders are a relatively uncomplicated illustration of how different "influences" came together through/in my
work. I could add influences from the conferences I attended and
the reaction to my performances both there and back in the department; the bits of writing I sent to academics outside Bristol and the
comments I received in return; the people concerned with the submission of my Kentucky M.A. which was still unfinished when I
started at Bristol; the people I met during my "field work" in Jamaica; the ways I have since taught parts of what I learned in the
process to undergraduate students in Lampeter; reactions to those
aspects of this process that I have attempted to get published; the
ways I have tried to make sense of this process to those who might
want to employ me; and, indeed, those involved in the rest of my
life, none of whom could be separated from this process. But I don't
have the space, here. I had much more space in my thesis. You could
try to get hold of it if you're that interested. But you might not be.
So, I'll make the point here that moving from place to place, making
connections with people in different places, negotiating the discourses and "opportunity structures" within and between such
places, and living out the contradictions that inevitably result from
this process is part and parcel of the life we all lead as we try to do research, to get things published, to gain qualifications, to carve out a

career path and, maybe, to make a difference to ourselves and to other people.

So, This Is the "Expanded Field" and You Write Your Way Through It Autobiographically?

Yes it is, and yes you can, but you'd probably have to justify this in "the literature." This is my experience, at least. And describing these connected spaces as an "expanded field" is not my idea. I've borrowed this from Cindi Katz (1994). Both she and, more recently, James Clifford (1997) have made much of the kind of account I've offered above. They've used this kind of thing to critique the dominant convention of "the field" being seen as a separate space in which research is done. They have pointed out the importance of the complex "spatial practices" of ethnographic research and of the politics of translation within and between the wide variety of locales that make a difference in such work. But there's also the related politics of knowledge that affects what kinds of research gets done, on or with what groups of people, in what kinds of places, by what kinds of researchers, from what kinds of places, for what kinds of reasons, and with what kinds of audience in mind. So an important thing to ask is who gets to "do" and who gets "done" in the expanded fields of ethnographic research? These arguments are addressed, to a certain extent, in those parts of the anthropology literature in which the idea of "traditional" single-locale "fieldwork" has been opened out through debates about reflexivity.

Here, researchers have been urged to write about their involvement in their own research because they may or should feel that they have to try to make sense of the tricky circumstances in which they studied before claiming to know anything about what they have studied. Here, I imagine that you'd agree that relationships and interpretations developed within any "field area" can never make sense in and of themselves, just then and there. Researchers' relationships and interpretations always stretch or leak out of that space and time, as do those of the people whose lives they study. So, one argument that can be made to justify a multilocale approach is that the modern world is organized and experienced through local accommodations of, and contributions to, distanciated social/cultural/economic/whatever relations. Any ethnography in and of the modern world, so the argument goes, must therefore reflect this mode of organization and experience through its research design. But calls for the multilocale ethnographies that could do this have not just been calls for studies of different groups of people in different locales as their lives are related to one another. George Marcus (1995), for instance, has called for work in which the multiple sites between which a research project might be stretched are not set out beforehand, but emerge through the process of following "emergent objects of study" and seeing what networks can be traced out in the process. These "objects," he has suggested, could be people, metaphors, conflicts, and/or things.

It now seems in this context that I had set out to follow a fruit thing and then found out that the locales connected through this process extended beyond the "thing system" I had intended to trace. I had started out trying to organize the first kind of multilocale research and had ended up with a project that was much more like the second. The worlds which were connected and, indeed, that blurred into each other through this process all had a bearing on what I was able to study, how I was able to study it, how I was able to (mis)understand this blurring and how I was able to represent and perform this (mis)understanding to others. Looking back on this, I concluded that my "emergent object of study"—the only "thing" that connected the multiple locales of my research—was "me" (whoever and whatever that might be). So, I could say that writing your "self" through the "expanded field" of your work can be justified "academically" as a way of writing reflexive, multilocale ethnography. I'd like to believe that this is "true." The literature that can allow me (at least) to think that I might not be alone in this belief is growing (with the help of writers like Clifford and Katz). The production and consumption of this book is, I hope part of that process. You are welcome to refer to this chapter if you think it might give your own writing some more credibility. It might not, of course. We'll wait for the reviews.

Is That the Reason That You Did It, Then?

Nope. Writing autobiography in and of this expanded field was not something that I chose because it might be interesting, because esteemed people were clamoring for it in "the literature," because it was the latest clever, "cutting edge," groovy thing to do. Things were much more desperate than that. In the third year of my Ph.D., I had found myself unwilling and unable to write up most of what I'd found out—a.k.a. the "data" I'd helped to construct—in Jamaica. This reluctance was caused in part by a crisis over the transcription of the interviews I had conducted with farm workers in an "English" that I could only partly understand. Here, I was stumped by the problem of how to understand and to represent what I'd (mis)understood in our discussions. I thought I'd found a way of getting around this problem by paying a linguist from the University of the West Indies at Mona to transcribe these tapes in Jamaican so that I could read them carefully and quote from them at length. She told me that some of my misunderstandings were so bizarre that she and her husband had rolled around on the floor laughing at them. I asked her to annotate the transcripts when this happened. And I could just about afford this specialist transcription. But the sudden devaluation of sterling on the "Black Wednesday" of autumn 1992—just after I'd returned to the United Kingdom—meant that the transcription costs of the tapes that remained with her in Jamaica had rocketed. I spent a great deal of time and effort desperately, and unsuccessfully, trying to find the extra cash to have them "done."

This financial shortfall was not the last of my worries, though. In the January of the following year I presented a paper about this research at the Institute of British Geography's annual conference. I'd been placed in a session entitled, "Bringing the 'Exotic' Back Home" with a paper entitled, "Constructing the Exotic: The Case of Tropical Fruit." The session drew a large audience to what was the first conference paper I had given in the United Kingdom. It was also the first and only time that I have tried to write a whole fruit-thing-system paper. I found it almost impossible. There was much to fit in and, in the paper I prepared to speak from, it really showed. I was going to read from a text that was about as long as this book chapter. The arguments were just as carefully made. I would have to read it word-for-word. But I had only the usual twenty minutes in which to present it. I was incredibly nervous and far too eager to be liked by the audience. So, in the presentation, I rose to the laughter that greeted some of the things I said and went off on some anecdotal tangents from the text. If I were a stand-up comic, I'd probably refer to this as "playing the audience." But I was a third-year Ph.D. student giving a talk about globalization, inequality, structural adjustment, the experience of deepening poverty, the strict supervision of labor, and the politics of race, gender, and class that ran through this topic—that kind of thing. I don't think that I even got halfway through what I wanted to say before I was asked to stop. Time was up. You might have been there. I hope not. I found the audience's reactions to my performance extremely disturbing both in the questions after the paper and then in the comments that rolled in after the session was over. Perhaps it was just me, though, who thought this was disturbing.

I had faced this audience as yet another white, middle-class, English male academic who had done his research in the "Third World." To those who were most offended, I seemed to have managed to present myself as someone shoring up the kind of white, middle-class, male English privilege that I thought I was undermining. I was told that I seemed to be totally oblivious of the power relations that had worked through what I had been able to do in my research. And along with many members of the audience, I was accused of finding racism "funny." I had screwed up, big time. And it was humiliating. But then, later, a number of people told me that the paper had been interesting or humorous or a bit of "light relief" in the midst of a typically dry and tedious conference. A woman whom I had never met before stopped me the next day and said something like, "I think of myself as a feminist, but after what that woman said to you yesterday, maybe I shouldn't be." Oh. A supportive comment was most welcome at the time. But this was a difficult one to respond to. That hadn't been my intention—to turn academic women (or men) away from feminism. Believe me. That's a frightening thought. Under the circumstances, perhaps "Oh" was the best response. But what made my anxiety much more acute was the fact that the criticisms were published soon afterwards in the IBG's *Women and Geography Study Group Newsletter* (Madge 1993; Superville 1993). A friend who had also done her Ph.D. research in Jamaica while I was there was in this study group. She phoned me to break the bad news. I wouldn't believe it but she'd prefer that she told me rather than someone else. She read it out on the phone and then sent me a copy in the post. I was horrified. For months afterwards, I asked myself how I could have seen myself as being antiracist, antisexist, and so on while others saw me as reproducing white racist discourses and male privilege through what I said and did? What was it about *me?* And what could I do about it? My supervisor advised me not to worry about it. You get used to being trashed at conferences. It happens to everyone. But I couldn't help worrying about it.

At this point, I'd prefer it if you didn't feel too sorry for me. Worse things have happened to plenty of people. And these criticisms were part of much larger debates, anyway. But, given the difficulties that I had had (and helped to create) throughout this Ph.D., this felt like a "final blow" to me. I was ready to "submit," but not in the sense that I had been "advised" to do often, that is, within four years. Picture yourself in these shoes, then. You have nine months of writing time left before your funding runs out. Only a tiny proportion of your data is in the kind of state that would mean that you could happily analyze and write about it. You have been rocked by the reception of your first attempt to make sense of this in a conference performance. You have managed to get a lecturing job, though, that will start when your funding runs out. Unfortunately, this work will take up most of the following year. This is the year whose end must see the submission of your Ph.D. "on time." But it doesn't really matter whether you meet this deadline as the university's submission deadline is another year away. And the fact that your late submission might push the department back on to the funding body's "blacklist" might not seem like a bad thing to you. But, of course, a late submission might just mean that your supervisors' initial concerns that you wouldn't finish on time would be proved right. That's the last thing that you want. You had refused their "advice" from the start to cut down your research to a single-locale study because that wasn't what you had applied to do. And you'd blamed much of your slow progress on their style of supervision, which you had openly criticized. So you decide that you will write the only Ph.D. that you can hand in "on time." This Ph.D. will explain why your proposed Ph.D. research couldn't be written up and handed in "on time." It will include an "autoethnography" of the process through which you struggled to do your Ph.D. This portion will include, but by no means be limited to, accounts of your "supervision." And it will also be the place where you try to write your way out of the "identity crisis" provoked at "that" conference. You'll try to do this by constructing an account of your childhood and adolescence in southwest England and then try to make a better sense of it through talking about "cannibals and missionaries," Tarzan books and films, the Scouting movement, and Stanley's "adventures" in Africa which were part of this youth. Maybe. And you'll conclude the resubmitted version, at least, by discussing an undergraduate course that you subsequently taught in a small Welsh university. I hope you'd never want to do, or have to do, anything like this. But the point here is that I felt that I did. Who on earth would *set out* to write a thesis anything like this from the start? Not I—that's for sure.

So, We Need to Be Careful We
Don't End Up Like You, Then?

In many senses, yes, but, in others, maybe not—see what you think. I've used this warning once before in the title of a talk I gave in the Easter break of 1994, six months before my four-year submission deadline. And I think it's apt here, too. Then and there, I had been invited by my (then remaining) supervisor and a fellow Bristol grad student to talk at a conference for first-year human geography Ph.D. students. The session that they were organizing was called "What happens when things go wrong?" As you might now appreciate, they had invited a speaker with considerable expertise in this area. I took this occasion as an opportunity to put together the arguments that would explain and justify what I planned to write up that summer. But it was also an example of how these arguments emerged out of, and were performed as part of, movements through the expanded field of my research. It's quoted in full in my Ph.D. thesis. There's not enough room for it here. But I'll try to conjure up some sense of what it was about, using the following words (and what you can read into them, of course).

Imagine yourself in a small banked lecture theatre in a manor house owned by a large university in the south of England and run as its rather plush conference center. You are a first-year human ge-

ography Ph.D. student and are in an audience of perhaps thirty people. Most of them are also at the same point in their research careers,
but some are lecturers in the departments that make up the consortium that has organized the conference. One might be your supervisor. You are about to be told about "things going wrong" by the three
speakers who are sat behind a table at the front. They've all just about
finished their Ph.D.'s. You would like to be in that position one day
(in less than four years' time, of course). Attending this conference is
part of your research "training." One of the speakers is a tall, white,
scruffy-looking, twenty-eight-year-old English male. He is the last
one to speak, and does so from a script in which what he says is apparently written word-for-word. He won't stand up, or even sit on
the table. He is really nervous. He'd had a rough ride with his last
conference presentation. He is reluctant to look up from his script as
he speaks. He grins a lot, but not because he thinks what he is saying
is particularly funny. Or at least that's how he remembers it four
years later, as he's writing a book chapter like this one.

He begins with an anecdote from his time as a grad student in
Bristol. In his third year there, a story had come his way of a fellow
grad who had been warned by a faculty member, "You want to be
careful you don't end up like Ian. He's all over the place." He says
that he's going to use this rather flippant point to address the issue of
"things going wrong" with Ph.D. research. He tries to situate his arguments in "the literature" within and beyond geography, where the
politics and ethics of ethnographic research have been discussed. He
suggests that, while you have probably been encouraged to research
and write about, through, and against the "dodgy politics" of the
world around you in your work, you probably haven't been encouraged to research and write about these politics closer to home, in
your department or university, for instance. He is about to do just
this, though. He doesn't believe in that "ivory tower"/"real world"
divide (and you probably don't believe in it either). He has plenty of
experience of academia as—in its own way—an exploitative and
cutthroat business. He'd love to have read James Sidaway's (1997)
paper about "The Production of British Geography" before preparing this part of the argument. It would have backed up his point really nicely, in "the literature." It's in *Transactions,* you know. But the
research for that paper was only in the early stages. Sidaway was at

this conference and he would drag Cook off that day—or was it the day before?—as one of the forty interviewees for his project. The scenarios that Cook wanted to outline in his talk to you for that graduate audience and to Sidaway in that interview weren't that un-usual. But "proper research" undertaken that close to home and published where it surely must demand respect hadn't come out yet. So he can't say that with much more than the authority of personal experience, whatever that is. He has to be a lot more speculative, provisional, suggestive. To you, he may not sound as if he knows ex-actly what he's talking about. But, there may be a spark or two of recognition as you catch various bits of what he's saying. Or, your mind might have wandered off, I suppose. Mine does.

He mentions how the transition from being a graduate student at the University of Kentucky to being one at the University of Bris-tol was difficult; how things went wrong "in the translation" (so to speak) between two very different academic locales; how, in the process, he became a rather bitter and angry student who was a pain in the bum to supervise; and how things got worse during his third year owing to a lack of funds and a disturbing reaction to "that" con-ference paper. But he tries to make sense of this process not by ref-erence to theories addressing the contemporary politics of education, but to a short passage taken from Michael Taussig's (1987) book *Shamanism, Colonialism and the Wild Man*. Here, he says, Taussig tried to make sense of the "cultures of colonialism" and the dynamics of their (trans)formation by South American people and the Europeans who colonized their territories at the end of the nineteenth century. And he quotes Taussig's argument that, there and then, there "were, in effect, new rituals, rites of conquest and colony formation, mystiques of race and power, little dramas of civ-ilization and savagery which did not mix or homogenize ingredi-ents from the two sides of the colonial divide but instead bound Indian understandings of white understandings of Indians to white understandings of Indian understandings of whites" (Taussig 1987, 109). He (Cook, not Taussig) then makes the point that, if you re-move the physical violence and communication problems, and then substitute "cultures of cleverness" for those of "colonialism," you might have a useful way of thinking about the politics of graduate student life in your department. It worked for him. Sort of. To make the point as clearly as he can, he paraphrases Taussig's argument by arguing himself that, on entering departments, first-year graduate students are initiated into "cultures of cleverness" that bind faculty understandings of graduate student understandings of faculty to graduate student understandings of faculty understandings of graduate students. The way that he has written his paper will illustrate this point, if you don't quite get it yet. He had to think about it for a long time before it clicked.

He then goes on to say that he takes the faculty/graduate student binary with a pinch of salt, that these aren't the only people whose understandings of one another's understandings help to create, sustain, and transform these "cultures," and that these relationships have to be seen as situated within the wider cultural politics and political economies of higher education. There are, for instance, still far too many white, middle-class men like him, and perhaps like you, too, teaching and researching in geography departments. This is certainly the case in the United Kingdom. He also knows exactly what feminist geographers like Linda McDowell mean when they criticize these academic "cultures" as being invariably "masculinist." He quotes one of the quotations that she uses to make her point. This describes graduate students' initiation into such "cultures" as learning the:

> process of one-upmanship by which we learn to be critical thinkers. In graduate school we are taught that the measure of our intelligence is the extent to which we can show others to be wrong. Thus the best students are those who can offer the most masterful critique, pointing to the methodological flaws, finding gaps in the argument, and using the most sophisticated language. One consequence is an enormous loss of self-confidence and self-esteem, so that it is the unusual student who emerges from a graduate program as a confident scholar who feels good about herself or himself. (Anderson [1992]in McDowell 1992a, 402)

This neatly encapsulates his experience of being a Ph.D. student, and it's nice to know it's not just a Bristol thing (it's all over the place) and he's not an unusual student. When he first read this passage it was, at once, a depressing and fantastic thing to be told. He thinks that this passage may have a similar effect on you. He can't possibly know, though. But he then goes on to talk about how the rules governing the allocation of funding in higher education have a huge effect on what any of you, and he, can do. He is on much safer ground, now. What he is saying is probably more obvious to more people. It's one of those important but boring things, to him at least.

He then tries to quicken the tempo. In one of his characteristically long sentences, he says that, through writing about graduate research as embedded within the kinds of tensions, contradictions, and inconsistencies that constitute these structures of power/ knowledge—through ambivalent complexes of accommodation and resistance, through contextual performances of identity and so on—he believes or hopes that you (or at least he) can capture, think about, and at least partially deal with what goes wrong with such work *in* such work. The point he's making is that, when you think that things are going "wrong" with your research, they might be going "right" if you think about them differently—as something that you can learn from and perhaps follow up. And he would like you to learn from his "mistakes" in this respect, if you can and want to do so. He argues that reading about the "expanded field" is useful training for these mental gymnastics and recommends that you read Katz's (1992) work in particular. By design and by accident, he believes that there's much to learn if you tackle head-on the "fact" that your research is almost inevitably going to go, and going to be, "all over the place." He seems to want to turn this criticism into an observation, or perhaps even into a compliment.

So, His Main Point Is That You Can Always Somehow Turn Your Shit Around?

Oh. I'm not sure, I wasn't really paying attention. But you should already have an idea about the theoretical positions that he cobbled together to make sense of his research from "the literatures on" reflexive/multilocale ethnography, cultural politics, political economy, and Taussig (wherever he fits in). Actually, he's probably "all over the place" as well. You want to be careful you don't end up like him, too. I hope he won't mind my saying this. But, maybe you should be careful that you don't end up like yourself, too. But—

going back to "me" talking about "me" rather than "me" talking about "you" (not) listening to "him" (Cook, not Taussig, that is)—I want to get serious now. Please concentrate. One of the things I was trying to piece together in this talk was a way of thinking about the ethics of my activities in/between this expanded field. But I didn't then have the confidence and, more importantly, hadn't done "the literature review" to tackle the issue of ethics head-on. This is a part of my research—then and since—that I've been reluctant to stake out for inspection by people I don't know, by people who might not give me the benefit of the doubt, by people who might want me to have a crystal-clear position. Where could this reluctance to talk for fear of being humiliated come from, you may wonder? Perhaps I've already told you. And where does this chapter fit into all of this? I'm sure that, by many people's standards, the research that I have done and the way that I have written about it could be labeled as "unethical."

Anyway, close to home, I had effectively conducted covert participant observation in the department where I did my Ph.D. research. And, even though I took the usual steps to throw readers off the scents that would clearly identify the key characters in this account, most human geographers (in the United Kingdom at least) wouldn't have to hire a private detective to know who was who. Closer to home, I wrote an account of my childhood and adolescence that, among other things, picked apart the everyday moralities through which my parents had tried to bring up their three boys to "do the right thing." Not many parents have to face up to such an analysis in print, and I hadn't asked their permission to do this one. And, further away from both of these "homes," I'd had to negotiate the ethical minefield of my "proper" research in and around London and, in particular, in and around the two fruit farms in Jamaica. Then, for instance, the question of what I should say to people in one part of the fruit system on the basis of what I'd learned from people in another was constantly on my mind. By accident and by design, there were many things that weren't, and weren't supposed to be, known by everyone involved. Yet, I was piecing this knowledge together for a thesis that could, eventually, be read and acted upon by some, more than by others. And I'd spilled some beans along the way. Power and knowledge were everywhere. So, the point here is that a number of ways of being "me" were (un)ethically experienced, challenged, performed, negotiated—take your pick—as I moved through this expanded field to do my research.

Did You Illustrate How You Dealt with This, Then?

Yes, in a small way. The presentation included a section where I talked about the way in which I'd responded to a challenge made by two white Jamaican men, one an estate owner and the other a farm manager. They were "friendly rivals" in the export agriculture business. Let's call the estate owner "Tim" and the farm manager "Jim." In the early stages of the farm research, I rented a room in Tim's "Great House" where the person who'd sent me the report about my conference presentation was also living while doing her Ph.D. research. I wanted to live there, too, so that I could cycle to the neighboring valley where Jim managed his fruit farm. Tim had helped me to gain Jim's consent to study what went on there. And later I had the opportunity to house-sit a property owned by Jim's brother and his wife in the grounds of his farm. I got to know Tim and Jim very well through my research, both as people from whom I learned a great deal about fruit farming in Jamaica and as people with whom I would hang out socially. But, as my research progressed there over a period of six months, the hospitality and frankness that they had initially offered became increasingly punctuated by their anger over the "brass-necked" nature of what I was doing. What, they argued, gave me the right to swan into their lives, look closely and critically at their finances, business methods, family lives, and, perhaps most sensitive, ways of dealing with their increasingly impoverished workforces, and then fly away and write about this as if I didn't equally owe my livelihood to the ugly means of exploitation I obviously saw in theirs? Since at that time, my parents had been running their own business for thirty-two years (it went bust in my first term at Lampeter), their most disturbing question concerned whether I would even *consider* researching how they had made their money off other people and then speak about it critically in an academic arena. And, although much of this line of argumentation could be seen as tactical—their playing off what they saw as my "misplaced socialist idealism" against what they knew about my family background to persuade me where my ultimate loyalties should perhaps lie—I could not deny that they had a point and if these ideals were to remain somehow intact, I would have to deal with it in my work.

Anyway, in the presentation, I dealt with this issue through talking about how I'd adopted an identity politics in which, among other things, you try to state exactly where you're coming from. Few people are going to be able to tell from just looking at you that you are working through, for instance, an antiracist standpoint. It's also possible that few people would be able to tell this from seeing you present a paper at a conference—particularly if it's one of those that go wrong. You can't assume that people will just assume your position, as you understand it, at least. If you can understand it. But it's not just a case of making a simple statement of where you think you're coming from. That's not the answer.

What Is the Answer, Then?

Twelve. Ha, ha. If only. No. This is where some of those mental gymnastics I mentioned earlier came in useful. At the conference, I had performed them with the help of Michael Taussig. Six months later, I had handed in my Ph.D. Exactly the same Taussig-like arguments were in there, too. But, after the *viva,* my examiners wanted more. They wanted me to contextualize the approach I'd taken in "the literature." What they'd read had been contextualized in the "expanded field" of the research that I'd tried to do, and what had been drawn into it along the way. They wanted a separate context, though. In the end, I was glad that they had "made" me do this context. It helped me to make better sense of what I'd done. It gave me some ideas about how I could do the contextualization more effectively the next time. Maybe. And their requirement highlighted and helped me to work through a fundamental problem. How on earth could I state exactly where I was coming from when I didn't believe that I, or anyone else, could come from "exactly" anywhere? I still feel that I had to make that statement. But I've also become even more convinced of the impossibility of the task through writing this chapter. Answering this question, as you can probably imagine, wasn't easy. But I think I managed it. And I found the work of a range of feminist and postcolonial writers invaluable here.[1] This work was discussed in detail in the introduction to my resubmitted thesis. It had to be. But, for this chapter, a brief summary will have to do. Have you heard about that theatre company that does thirty-second Shakespeare plays? How about a thirty-second literature review? Here goes.

There are three major narrative conventions in autobiographical writing: (a) the "great man" who spins out his neat, progressive, linear, "public" life; (b) the "confessional," where he recounts how "private," "behind the scenes" dramas knock him off his course; and (c) the "bildungsroman" where he combines these in tales of triumph over adversity. "He" is usually, but not always, a certain kind of man. Although these conventions have their differences, they are united in the assumptions that they make about the self. It is an essential, unique, rational, and bounded entity. It is called the "sovereign individual." And "it" possesses all of the stereotypical characteristics of the white, Western, middle-class, heterosexual, nondisabled male. "He" knows exactly where he's come from, where he's gone and how he got there. "He" is a discrete, rational, coherent, progressive, and whole person. "He" may "fall to pieces," but he can pull himself together. "He" can do this "heroically," too. But "he" is also pulling a big power trip by writing about his life in this way. "He" has the luxury of not being reminded every day that he is "different" in terms of his skin color, his nationality, his class position, his sexuality, his bodily status, and his gender. "He" is the norm, after all. He has the luxury of being "unmarked" by these things. They don't complicate "his" sense of self. But they do complicate everyone else's, in all kinds of ways.

But how do these more complicated people write about their lives? Can they do this through the narrative conventions that fit "his" life? Well, no, they can't, because these are patriarchal discourses, colonial discourses, ableist discourses, that kind of thing.

1. This includes, for instance, Stanford Friedman 1988; Anzaldúa 1990; Mohanty, Russo, and Torres 1991; Okely and Callaway 1992; Smith and Watson 1992; Visweswaran 1994; Behar and Gordon 1995; and Thompson and Tyagi 1996.

There are plenty of other ways of living a life and of (not) writing about it. But they don't get so much respect. People often think they're a bit odd, or that the people who write them must be a bit odd. They can look like streams of consciousness. They can blur spaces and times, fantasies and "realities." They can be written as if people are totally inseparable from other people, things, and processes. They can also be written as if these people's lives are fragmented, disconnected, by no means linear, or "progressive." They can be written in different voices, different languages, or languages that don't have an official written form. They can be written colloquially, as the things they describe might be discussed on the bus, in a conference bar, wherever. They can be playful. They can be serious. They can be both—often at the same time. Humor can bite, you know. And they can be written by people who want and need to claim a voice; who want and need to make connections with others: for pressing political purposes. People live their lives in so many ways. They are (un)able and/or (don't) want to write about their lives for so many reasons. But, as well as developing new ways of expressing their lives, they can also draw on those three older, major conventions discussed earlier. They could be doing this because these are the only ways they know a "self" can be written. Or they could do this because they want to use these conventions strategically, ironically, or in obvious juxtaposition, to show how daft and dangerous they can be. So, just because a writer employs them, it doesn't mean that s/he necessarily accepts the baggage that's attached to them. And, because the alternatives are often less straightforward, more tangential, more ambiguous, more "open," readers can make a lot more out of them. Writers can encourage their readers to be more creatively involved in making meanings out of their words. Readers are supposed to go off on their own imaginative journeys when reading these texts and, perhaps, well afterward. Writing should be about giving readers something to think with, not just to think about, or to learn, to master, to remember for an exam. Autobiography can allow the construction of accounts of the self that tell far more than the story of a single person, something like that. The thirty seconds are up. That's the review. I hope it made some sense.

Are You That "He" Man, Then?

That's a tricky one. I do have all of "his" outward characteristics. You might think I was "he" if you saw me present a conference paper. Especially "that" one. And you could read almost all of my Ph.D. thesis as "confessional." It "reveals" behind-the-scenes dramas in Bristol's geography department. And there are lots of "personal" things in there about growing up, too. I really exposed myself there. And I'm a bit embarrassed about it all. A lot of people have told me it was a "brave" thing to do, writing that kind of Ph.D. Desperate is the word I prefer to use. As I said earlier, I didn't set out to write an autobiographical Ph.D. It was supposed to be about a fruit. Remember? I couldn't do that in the time that was available. I used autobiography to write myself out of an awkward corner. I had to "pull myself together" to do that. And I had written my Ph.D. as a "sovereign individual," too. It was all my own work, and mine alone. No one else was responsible for the contents of that thesis. Oh, no. I had to sign something to that effect. Didn't I? Last year, two Lampeter undergraduates asked me if they could register to do a Ph.D. together. Now there's a thought. And, the fact that I have been awarded a Ph.D. makes this chapter a classic "triumph over adversity" story. I got the letter about two weeks after I started writing this chapter (although it's a 1997 Ph.D. on the books) (Cook 1997). Sorry I didn't mention that earlier. So, you could say that my thesis is "confessional" autobiography. And the fact that I got a Ph.D. out of it, and this has been "revealed" to you here, makes this chapter a classic "bildungsroman" narrative. Or does it? What do you think? How do you think this chapter would fit into the arguments in that thirty-second review? Why do you think it is written this way? How does it "work?" Does it "work" at all? Can you take it as a serious piece of academic writing? What is it about? Is it about the "true life story" of an important geographer? I think not.

When I was writing the first draft of that Ph.D. thesis, I hadn't read that literature. But I had a lot of those arguments in mind when I wrote the first draft of this chapter (Cook 1998). It's a lot longer than what you've got here. And it goes a lot further off the beaten track. It does not "reveal" me as a cool, calm, logical, organized, strategic, linear-thinking person. I don't think this version does, either. I'm not like that. And doing my Ph.D. certainly complicated my sense of self. I was "all over the place." That Bristol lecturer was right—well spotted. But, my Dad read the first draft. His main comment was that this was a family trait—on my Mum's side. She has a "butterfly mind." So did her mum. My gran. And my Uncle Gordon. He was a policeman, you know. I think a butterfly mind is quite handy in my line of work. Maybe not in his. It's geographical. Handy. Aren't we supposed to see and make connections between people and the worlds they inhabit? Between spaces and times? Between everyday life and larger social/cultural/economic/political/et cetera processes? Between travels, translations, and transculturations? Between theory and practice? Between plenty more? Across all those categorizations, boundaries, and borders? In complex ways? And couldn't autobiography be a useful way of doing this? Not the only means, I hasten to add. The people who organize graduate student conferences like the one that tall, scruffy geographer spoke at certainly think so. People who have faced dilemmas in their Ph.D. research, who have learned from them, gotten through them, and can talk about them are valued presenters. With this context in mind, I'm writing something I would love to have read at that stage in my academic career. But it's not just a methods thing. Is it? For me at least, it's also been an extremely liberating way of writing about what I've learned from my research—what I really should be writing about, perhaps (i.e., not "me"). It may sound odd, but I was incredibly excited after I finished the first draft of this chapter. That's not my usual reaction to finishing off an academic paper, believe me. Most of that kind of writing is very rigid, ordered, linear, progressive, boxed in. Isn't it? And, I wonder, do people really think that way? Are they really like that? Is that really how and why they did that work? Wasn't there any ambiguity, surprise, confusion, uncertainty, emotion, irony, pain, leakiness, life in what they did? If so, where did it go? Why didn't it get onto the page? Couldn't we learn something from reading about these things? I have wondered whether there was something wrong with me, in this respect. I think others have, too. I enjoy reading about those things. I struggle to represent my research in that "proper" way—wearing that straitjacket. It has made me wonder whether I'm "clever" enough for this game. I find it very alien, although not as much as I used to. I'm fairly sure I'm not alone there. But it's a performance. Isn't it? It is, as auto-biography is a performance; as this chapter is. But is it a performance in the same way? What have you "learned" from reading it? Not much about "me," I hope. It's not a me-me-me-me-me-me-me-type narrative. Is it? I think it's an it-me-them-you-here-me-that-you-there-her-us-then-so- . . . narrative. It's an "expanded field" thing. And you're in it too. Aren't you?

A Self-Reflective Exploration into Development Research

Robin Roth

I engage in discussions about postcolonialism, and I want to learn more. I see the media blaming environmental problems on the poorest countries, and I get upset. I read about development projects with dreadful social and environmental consequences, and I get angry. I read critical development theory linking the social, the environmental, the political, and the economic, and I get excited. Yes, I *actually* get an adrenaline rush. I become a cliché; I want to "get involved."

In other words, I have a passion for international development issues. I love the complexity of how global and local economies, politics, cultures, and natural environments intersect in "Development work."[1] I'm entering graduate school now and want to pursue this

Thank-yous to my father for his cherished influence, and my mother for her continued guidance. I would also like to thank Geoff Whitehall, Michael Keeling, and Alice Hovorka for their comments and helpful suggestions. I would especially like to thank Pamela Moss for her continual support and careful editing of this piece. Any errors or oversights remain my own.

1. I recognize that the term "Development work" is problematic because it assumes both that there is work to be "done" by someone from the outside and that the work will result in a linear change for the better. Also implicit is the notion that the work needs to be both initiated from and completed by the outside in order to maintain the linear progression of "development." I do not wish to imply this by the use of the phrase, but only use it for lack of a better one. I define "Development work" as the process by which a researcher displaces him/herself to a community in the Majority World in order to conduct research in the area. Implicit in this is a desire to gain a better understanding of the conditions in the Majority World and per-

passion. Yet, I have an uncomfortable feeling about doing so. I'm concerned about conventional Development discourse and praxis which promote a (continuing) colonial[2] relationship between Minority and Majority Worlds.[3] I'm also concerned about the tendency of Development practitioners to privilege conventional Minority World knowledge.[4] I don't want to perpetuate this tradition; I want to help create a new one. My friends question my motives, "Oh, you want to save the world—what a noble cause!" I'm not noble. I want to *do* Development work because I *like* it. I'm not selfless, gracious, benevolent, altruistic, or sympathetic; I'm simply interested. Or, at least, I think I am. I haven't "done" Development work yet. I still have to determine what my role will be; where I best fit; even *if* I fit. Despite my misgivings, and because of my passion, I'm anxious about acting on it.

I fear becoming what I criticize. I'm scared to withdraw from the hegemony in which I grew up. I'm scared I won't be able to withdraw from the hegemony in which I grew up. I worry that I'll choose the wrong path, that I'll spend years with a small organization and still feel no one is listening. Or, I'll spend years in a large institution and feel everyone is listening to the wrong story. I worry I'll mess up, that I won't be able to meet my own expectations, that

---

haps facilitate change. I choose to capitalize "Development" following the usage proposed by Irene Gendzier (1985), and followed by others (e.g., Escobar 1991), to highlight the historical, "invented" nature of this discourse.

2. The term "colonizing" has come to signify more than the military domination of one country by another. It is also used to refer to the appropriation and suppression of another's culture, belief, and lifestyle. For the purposes of this paper, I define colonization as the process of mapping the Other with a "discursive or political suppression of the heterogeneity of the subject(s) in question" (Mohanty 1994, 196).

3. I am aware of the problematization of the terms "Third World" and "First World" and have therefore chosen to use the terms "Majority World" and "Minority World." I do not intend the terms to refer to geographically bounded countries. I use them with the understanding that the conditions and positioning they represent can exist within the same country, even within the same city.

4. "Conventional Minority World knowledge" refers to that body of knowledge generated by people in the Minority World that does not challenge, but instead adheres to and supports, dominant assumptions rooted in rationalistic and Enlightenment thought.

I'll be part of creating a mess even worse than the one I wanted to change. When thinking about my future, I return to the notion that my desire to plan inevitably collides with my inability to envision a clear path for myself. When I contrast who I am and what I want out of my life—both personally and politically—with what Development is and has been, I wonder whether I fit.

When I contemplate a career engaging in Development work, my apprehensions intersect with a number of issues in Development discourse. Obviously, I hesitate to sort through all these intersections here. Instead, I focus on addressing my most immediate concerns: how can I act as I am—positioned along multiple axes of power, a member of the Minority World, bearing my own specific agenda—without promoting the continuing colonial relationship between Majority and Minority Worlds? How can I justify my participation in Development work? And how can I participate in a way that helps create noncolonizing spaces and produce noncolonizing relationships between myself and Others when our identities and experiences are so different from one another?

This analytical trip has a distinctly autobiographical component. I wrote the first draft of what was to become this chapter in the spring of 1997, three weeks before my dad died. This is important because the more I reflect on what I have written here, the more I realize that the very analysis held in these pages is not "objective" (if there is such a thing) but is very much part of my own autobiography. This chapter reflects my informal education and my formal training, both of which taught me to be critical and self-reflective. Politics, the environment, even human spirituality were familiar dinner-table topics to me by age fifteen. My dad and I used to sit around the dinner table and analyze the world and our place in it. Both my parents encouraged me to pursue, as they had, a diverse education with a strong component of critical thinking. It is no surprise, then, that I continue the family tradition here; what I think is part of who I am.

My thoughts here reflect my current situation as a graduate student. This chapter represents a moment in my life when analysis of, and reflection on, my present and potential position in Development discourse is both inescapable and necessary. Academically, I must sort through these issues as part of my training to be a researcher. Personally, I must forge a space where I can be who I am without losing sight of other people. I struggle with the incompatibility of my own values with those of dominant Development discourse. I struggle with my desire to "do" development work in a world where capital accumulation is valued more than the environment. I struggle with my past in light of my future. Once I act upon what I've written in these pages, my life will be shaped by what I think; who I am is part of what I think.

The following, then, reflects the mutually constitutive relationship between what I think and my life, between this text and my autobiography. First, I explore how Development discourse has been colonizing, and I confront my own potential to colonize. Second, I reflect on my own desire to be involved, distinguish the terms of my participation, and build a justification for conducting research in the Majority World. To this end, I consider a strategy Jennifer Robinson (1994) suggests for creating a noncolonizing research space. Third, I think through a conceptualization of a lesser-colonizing research space that addresses some of my concerns. I conclude by identifying potential challenges inherent in applying these conceptualizations in the field.

Development and Its Discourse as Colonizing

My interest in the colonial, and my distaste for it, has its origins in my childhood. My parents, who were involved in and deeply moved by the antiwar movement, instilled in me a dislike for unequal power relations, particularly as they manifest themselves internationally. I relate this to my current interests by taking the time to review some of the ways conventional Development discourse is colonizing. I must continually be aware of past practices in order to contribute effectively to the creation of alternative future practices and in order to shape my role in them.

Everywhere in the popular press, and often in the academic press, *they,* the Majority World, are compared to *us,* the Minority World. *They* are poor (and can't buy *our* products), *they* don't have adequate sanitation facilities (and are therefore unhealthy), *they* need better transportation networks (in order to facilitate *our* tourism). For the most part, practitioners from the Minority World identify the "problems" of the Majority World and suggest solutions that boil down to pretty much the same thing—*they* need to be more like *us*. The us/them distinction prevalent in Development discourse shapes the paternalistic and imposing nature of Development praxis and constructs the Majority World as an object to act upon. Continuing employment of this distinction leads to practices just as destructive as, and often very similar to, colonial ones, such as restructuring economies and cultures to suit the vision of the Minority World (Escobar 1995). Such discourse is just as colonizing as the more overt material practices of settling a "new" land.

Development discourse, then, it seems, has been created and dominated by conventional hegemonic knowledge about the Majority World. In other words, subjugated peoples and environments of the Majority World have been organized, described, and defined only in relation to the Minority World. For example, Majority World women have been characterized as tradition-bound, poor, religious, domesticated, and victimized, whereas the self-portrayal of the "average" Minority World woman includes being educated, modern, in control, and free (Mohanty 1994). Third World environments have been described either as bountiful and rich (ripe for the picking) or exploited and damaged (ready to be saved), whichever serves better the interests of the Minority World.

Development discourse also privileges knowledge generated in the Minority World over that generated in the Majority World. Many development initiatives, for example, assisted in displacing traditional agricultural systems in favor of large-scale, industrial cash-cropping. The underlying assumption motivating these projects is that the Minority World has much to teach with nothing to learn (Sen and Grown 1987; Braidiotti et al. 1994).

Not only have development practitioners neglected to seek out Majority World knowledge but they have actively ignored and suppressed it. Many global citizens find themselves without forum "in which their voices are heard, the wrongs and harms they have suffered recognized, their claims validated and their ongoing struggles for justice legitimized" (Mendlovitz 1998, 110). Recently, however, the holders of Majority World knowledge have been forming NGOs, such as the International People's Tribunal, and are using them to speak—insisting their voices be heard. Now, more than ever, the colonial nature of Development discourse is being challenged by Majority World citizens and like-minded Minority World citizens who are voicing alternatives to hegemonic viewpoints.

There is little doubt that conventional Development discourse has actively contributed to the colonization of the Majority World. There is also little doubt that I have been complicit in this colonization. I have been almost wholly educated from Minority World sources and I am thus very much entrenched in the very hegemony I wish to displace. I recognize that I'm complicit in the ongoing power differential between the Majority and Minority Worlds, and that I'm also positioned as a colonizer. It is part of my autobiography. The question now is whether I can shape my future involvement so as to break with colonial Development discourse and help voice alternatives.

Becoming Involved in Development

Whenever I contemplate possibly participating in Development work I am constantly aware of its colonial traditions and consequently worry about my ability to overcome them. I recognize that my geographical location makes me both globally privileged and part of a hegemony that gives precedence to conventional Minority World knowledge. As a result, I must entertain the possibility that the parameters within which I am positioned place me in a potentially colonizing space. My unintended, but nevertheless real, autobiography puts me at risk of associating myself with the long and sorrowful tradition of white, middle-class researchers from the Minority World going to study Others in a Majority World country and categorizing and defining them according to a Minority World "norm."

In the rest of this section, I reflect on how I became interested in glocal Development and how these roots now shape the terms of my potential involvement.[5] I articulate what I hope to accomplish and how I hope to go about accomplishing it. This process exposes my previously articulated concern about pursuing a career in De-

5. I use the term "glocal" to signify that no process or issue is purely global or purely local, but manifests itself at numerous spatial scales.

velopment, which in turn leads me to question how to justify my participation in Development work.

The Roots of My Interest

As I begin engaging in Development work, I know that I have to act as I am. I need to feel good about who I am and what I am doing in order to be effective. This means recognizing the influence of my own history and my own positioning, without contravening my own value system.

Alongside political dinner-table conversations, my parents encouraged me to act on my convictions. I first heeded their advice when, at the age of fifteen, I helped organize a youth environmental group in my home town. Since then, the deteriorating glocal environment has been my primary concern. At that time, I began hearing that increasing human population was the biggest threat to the natural environment (all *those* people wanting cars of their own) and that "population control" needs to be the focus of the Development community. The second time I heeded my parents advice was at age nineteen. I became involved in the Vancouver Island Public Interest Research Group (VIPIRG), where I met people who, through their astute awareness of power relations, helped me realize the bias of the population control argument. Sure, numbers of people are important, but what and how they consume is what translates into environmental damage. Both the quality of human relationships to the planet, not only individually but also collectively, and the glocal impacts of human economic and political systems constitute environmental health. The population control argument is biased because it ignores the impact of privileged lifestyles on the planet. It is a one-dimensional view that focuses narrowly on population as the source of "problems" while neglecting other factors such as resource distribution, gender relations, and social support networks. The population argument is but one example of the frequent blaming of the Majority World for looming environmental disaster. This blaming reinforces the power imbalance between Majority and Minority Worlds that exists at various spatial scales and along axes of power. When I recognized this reinforcement, I became convinced that impending environmental crises need to be situated within a set of multiple and intersecting environmental, social, economic, and political processes.

The complicated specificity of a Majority World locale generates much of my enthusiasm for and interest in glocal Development issues. The intersection of various axes of power with glocal economic, political, social systems, and natural environments creates a situated specificity in which Development occurs. The material practices of conventional Development discourse rarely recognize these intersections. Instead, practitioners treat a locale or site in a bounded and fragmented way, not only denying the linkages and intersections present, but also homogenizing and simplifying the experiences of local people. Conventional Development discourse also appears to promote and enforce inequalities amongst local residents and between Majority and Minority Worlds.[6]

My desire to see practitioners of Development discourse and praxis recognize the complexity of Development issues shapes my personal/political agenda. With this agenda, to claim to be a neutral participant in Development practice would be to deny who I am—my own motivations, my history, my positioning, my autobiography. I need to act according to my own value system, which means I need to decide on the terms of my participation. I wish to participate in Development work that builds on what is honorable about conventional Development practice and that assists the creation of a new, less colonizing, Development tradition. I want to participate in research that explores the dimensions of a locale or concern, and I wish to do so while working with local residents, while not 'Othering' them.

I am also interested in global processes of power and how they manifest themselves at various scales. My agenda is certainly shaped by a desire to see a more pluralistic society where power is shared more equally amongst nations, classes, genders, and ethnicities, and between human and nonhuman nature. I'm certain that I will choose to be involved in Development initiatives that work toward

6. It is widely acknowledged that economic inequalities between the Majority and Minority Worlds and amongst genders and classes at the national scale are growing. See UNDP 1996; and Korten 1995.

decreasing inequalities of power while recognizing the intercon-
nections amongst oppressions.

In rereading this section, I'm challenged to confront what I read. I
sound as though I want my agenda to take precedence and that I
want to control what I do and how I do it, perhaps in complete neg-
ligence of local desires. I sound pretty colonizing. This is where I
often get stuck and it leads me to my primary concern—how can I
act as I am, positioned along multiple axes of power, a member of the
Minority World, bearing a specific agenda, without promoting the
continued colonial relationship between Majority and Minority
Worlds? If I conduct research while acting only on my own per-
sonal/political agenda, I would effectively colonize the people with
whom I would carry out research (and be complicit in maintaining
and reproducing colonizing ties). If I conduct research, as I am
tempted, while acting only on Other agendas, I would deny myself.
In denying myself, I not only place myself in an uncomfortable space,
but I also obscure my inherent role in constructing and shaping the
research. I would be fooling myself. I could easily talk in circles here
but these two polarized options obstruct what might be the best solution: to choose sites and people whose agendas I embrace, whose
agendas likely inspire my own to shift or expand. This melding of
agendas is preferable to the positioning, and presuming, of agendas as
conflicting. But even then, I would be idealistic to expect no tension
amongst agendas. Certainly, there are residents in a locale who may
have different interests than those of the residents I am working with.
Not only will there likely be a variety of not always harmonious local
interests, but there will also be the funding agency's interests and
those that come with participating in academia. Given that my desired involvement in global Development is riddled with concerns, I
must ask the question—is my potential involvement in Development work justified on this basis or is a noncolonizing space possible? In other words, how will my future autobiography unfold?

The Conceptualization of a Noncolonizing Space

When I realized that colonization could be subtle and that it
didn't necessarily involve the use of military maneuvers, I also real-

ized that individuals like me, could colonize. I can colonize by privileging myself over Others, by objectifying Others, and by attempting to control representations of Others. I can colonize by
promoting my own way of looking at things without attempting to
see what Others see. I can colonize by promoting my own ways of
knowing without attempting to know what Others know, and how
they know.

I was made aware of my potential to colonize while trying to
work across difference as an activist with VIPIRG. As a white, heterosexual woman I was challenged to find a role for myself, and a
means of interacting with others, that allowed me to support queer
and antiracist movements while simultaneously recognizing the matrix of power relations in which I was embedded. Working across
difference and across power differentials requires a strategy to avoid
colonizing. I need to be truly open, to engage with the people with
whom I carry out research, to allow myself to be changed by another, and to help create a space where exchange can happen freely
and not be based on unequal power relations. I need to find a promising means of accomplishing this before I can step into the world of
Development.

Robinson (1994) works toward this vision of a noncolonizing
space. She proposes the exploration of the space between self and
Other, between the researched and the researcher.[7] She incorporates
both positionality and voice into her conceptualization of a research
method that does not reproduce colonial/apartheid mappings and
placings of earlier researchers but that recognizes the presence of the

---

7. Up until this point I have avoided the use of the phrase "the researched" and
have used instead somewhat awkward phrases describing the relationship between
myself and the people with whom I would carry out research. I have chosen not to
use "the researched" because it reinforces the objectification aspect of research from
which I wish to distance myself. My choice is in contrast to Susan Hanson's (1997,
122); she states: "I have had an aversion to using the term 'researched,' precisely because of the distancing and objectifying pall it throws over those with whom researchers participate in the creation of knowledge. As Gilbert [1994] has
persuasively argued, however, some degree of distancing and objectifying is probably inevitable; in its jarring associations, the term, I have come to see, is actually
apt." I have introduced "the researched" here while discussing Robinson 1994 in
order to be consistent with her usage.

researcher and the researched. Building upon Gayatri Spivak's (1994) problematization of hegemonic discourse not recognizing subaltern modes of speaking, Robinson conceives of a space where the subaltern can speak, and be heard. She proposes, not to map Other, but to explore the space between self and Other where mutual experiences interact. Robinson suggests that researchers working with subjugated groups need to probe their positionality and acknowledge the implications of this positionality. She envisions mapping a space of in-between where the constructed meanings of both the researcher and the researched "could intersect, negotiate interpretations, explore contradictions and learn mutually" (Robinson 1994, 219).

I like Robinson's strategy quite a bit. She doesn't promote the complete objective picture traditional research strives for, but instead promotes a subjective picture that allows a space for subjugated perspectives and voices. She does not speak for and about the researched, but instead engages with them, recognizing their agency and allowing the construction and definition of the researched to be a joint venture. Robinson makes "the mediations of meaning and the interactions of interpretations among those engaged in the research process the object of investigation" (Robinson 1994, 220) and in doing so avoids speaking wholly from her place or in the place of another. Inherent in Robinson's proposal is an acceptance of difference and a desire to recognize this difference through the negotiation of partial meanings. When Robinson reports and thus legitimizes partial accounts, she is placing limitations on her own authority and helping to de-objectify the researched. And when she opens up to the influence of the researched, and hears them speak, she is actively lessening the traditional power differential between researcher and researched. Robinson's strategy allows her to be less colonizing than the traditional researcher.

I am not, however, at ease with everything Robinson proposes. She recognizes that one of the weaknesses in her proposal is that "the informant would need to be articulate and would inhabit a different discursive universe" (Robinson 1994, 221) and perhaps not engage in the same way, and thus the power relationship between the researched and the researcher would remain strong. The researcher must *also* give up control of the space and representations of it in order for this vision of a space to work. Otherwise, issues of hegemonic discourse structuring when and how the subjugated can speak remain unresolved. My other concern is that while Robinson's mapping the space of in-between recognizes personal positioning, it does not explicitly recognize that the researcher is also interacting as a representative of his/her culture and social positioning, nor does it recognize the relationship between this positioning and the production of knowledge. Robinson therefore does not take into consideration the external power relations that shape the space of in-between. Without a continual self-reflexive process in which the researcher is aware of these power relations the space could *never* be shared equally.

This strategy provides a framework from which I can begin asking unaddressed questions that I feel are important. Why conduct this research? Who benefits from the research? Am I not still mapping Other? Is it possible, and desirable, for everyone to benefit from participating in the research process? These questions, however, raise critical issues regarding control over the research in the early, planning stages and remind me that a thorough commitment to helping create a noncolonizing space is necessary from the beginning of the research process. The questions are also problematic because the very act of researching someone other than yourself, or some community, will be some form of mapping Other. Even, as Robinson suggests, the exploration of spaces in which mapping occurs is still a strategy designed to acquire knowledge, however partial, about someone other than the researcher's self. I do not think it is possible to know someone without "othering" or colonizing her. If this is so, then acting on my desire to be involved in Development work will effectively colonize. What remains then is my own question of whether I am justified in pursuing my passion in this way or whether I should simply refrain from participating in Development work and act on some other passion of mine.

There are a number of common arguments for and against doing Development work and several of them have entered my mind at one time or another. My dad used to present an argument that always struck a chord—"the work is here, not there" he would say, pointing to a copy of Korten's (1995) *When Corporations Rule the World*. I agree. The institutions and lifestyles of the Minority World need to be altered for the people of the Majority World to be able to determine their own path and I hope people work at that. But acting on my passion includes going elsewhere. Much of what excites me about Development work is the experience that comes with displacing myself into a different context and then exploring that context. The throwing into confusion of my own assumptions and expectations and the forming of new ones is autobiography in action. Besides, more and more these days "here" is "there," and "there" is inevitably "here."

On the surface, it appears as though colonization occurs with my displacement to the Majority World to do research. It is with this displacement that I colonize Others intellectually and concretely— it is inevitable that I, as a white, middle-class, heterosexual woman from elsewhere, will have an influence on the community I study. At the very least I will add a dimension of power that was not concretely present before. This is so even if I try to speak from a space of in-between. Is refraining from displacing myself the solution?

Global power relations help shape my positioning in relation to Others regardless of my geographical position. This cannot be ignored whether, for example, I am abroad or in Canada. If I remain in Canada, I am privileged in a global sense; I drink coffee, eat fruit, buy tennis shoes. Thus my presence serves to perpetuate the Minority World's continual economic and cultural colonization of the Majority World. If I go, I am in a position of privilege both in local and global senses and my presence and task of research colonize the people and environments with whom I engage. There is one substantial difference between the two situations, however—when in Canada there is little opportunity for a dialogue between myself and those I am colonizing, whereas when participating in Development work there is an opportunity to construct the research so that the community has an impact on me and the work I am conducting. There is also the possibility that my research will work to reveal and destabilize those discursive and political power relations that help to subjugate the inhabitants of the Majority World. It seems to me that my influence over the Majority World is a product of the society in which I live and that it is up to me how I, actively or passively, shape (or destabilize) that influence. So, with this in mind I think I can justify participating in Development work as well as in a new nonconventional Development discourse. Since I believe a truly *noncoloniz-ing* space is difficult, if not impossible, I now turn to questioning how to create these opportunities for dialogue and the creation of a less oppressive space.

Toward a Less Oppressive Space

In this section I work through how I envision a less oppressive space and the difficulties I might have in applying it. This is where I take my analysis and allow it to shape my future role in Develop-ment; allow it to shape my future autobiography. I base this concep-tualization on my own criteria as they have evolved in this paper, and it suits me, possibly only me. I think that how one plans to go about research is necessarily personal. My design of these criteria al-leviates some of my concerns regarding the nature of Development and its compatibility with my personal/political agenda.

I see all space as having the potential to be oppressive and the po-tential to be nonoppressive. I envision the multidimensional space between myself and another as the site where identities, oppressions, historical and political contexts, and power relations interact; where relationships exist. The same space exists between Minority and Majority Worlds, and it is here where researchers interact with the people with whom they conduct research. This space holds aspects of both identities and realities so that both I and the people who are part of the research are comfortable and have a voice—as long as this space remains unclaimed then it can be nonoppressive. As soon as someone lays claim to that space in an attempt to control it, the space becomes oppressive. There are four strategies I need to employ in order to help create the least oppressive space possible while still adhering to my personal values. First, I need to search out the peo-ple and sites whose agendas either closely match my own or inspire me to shift or expand my own personal/political agenda. This strat-egy lessens the possibility of a conflict amongst agendas and ensures a more cooperative research process. Second, I need to respect the space between myself and the people with whom I carry out re-search. This means not privileging myself, not silencing, nor speak-ing for those I am working with. Third, I need to recognize the multiple historical, environmental, political, cultural, and economic contexts within which the space is situated. This strategy requires me to have a broad perspective that recognizes the complexity of a situated locale and the interaction amongst external and internal factors working to create that specificity. Finally, I need to position myself in a way that is culturally intelligible and that actively reduces any of my perceived authority. This last strategy, in particular, requires me to have a practical and experiential working knowledge of the culture in the research because authority manifests itself differently in different places. But these strategies are a mere starting point—they are what I must commit to in order to maintain less oppressive relationships. With these things in mind I can more effectively help, as Robinson suggests, explore the space between myself and another where our mutual experiences interact.

This conception makes sense to me, but can this, or any other conception of a noncolonizing or less oppressive space, be applied? I venture to say that this approach to research is difficult to implement and provides me with many challenges as I contemplate participating in Development work. As a researcher I operate within a web of power relations. Although there would be room for me to maneuver, for example, in setting a research methodology and in choosing subject matter, often control over a project would not be entirely mine to give up. External power relations also shape the space of in-between. As I participate in Development work and contribute to Development discourse I have to ensure that I relinquish all authority over the research possible, including any textual representation, while maintaining my own set of values. My challenge will be to engage in a continually self-reflective process in which I recognize those multiple historical, cultural, and economic contexts within which, not only the community, but also I, am situated.

I recently had an experience working in a Majority World village, and I now recognize that regardless of how much I plan ahead, think through issues, read theory—everything is more complex while one is in the field. Upon my arrival the locals perceived me, particularly the employees of the woman with whom I was living, as having authority. As time went on I actively tried to reduce my authority, by helping out and being friendly with the woman who cleaned the house and with the woman who cooked. It worked; my authority was much reduced and I was no longer considered high up in the power structure. Yet I was also considered unimportant. I realize that if I were conducting my own research there, this loss in authority and importance would have made it difficult for me because I would not be taken seriously. I didn't *act* as if I were an important researcher and therefore I wasn't treated as one. I was not culturally intelligible in the way I wanted to be and, what's more, after six months in the society I am still unsure how I would maneuver myself so that I could be. Just getting used to how work gets done, and adjusting to different ways of measuring my success, was a long process.

This experience demonstrates to me just how complicated social and cultural systems are when you come from outside, and just how difficult it is to conduct research in them. I understand now why many researchers decide to dedicate their research careers to one region of a country. The experience also taught me that I need to take my theory and my thoughts on "how to" and try them out. I believe I have something to contribute to a re-created Development discourse but nothing will be easy or perfectly planned. I will make mistakes, both in my eyes and in the eyes of the people I will work with. But, through mutual learning, I will also learn how to correct them.

My own autobiography is at a turning point. In this chapter I have analyzed my position vis-à-vis dominant Development discourse and attempted to frame an approach to my future involvement in the world of Development. This analysis has come from somewhere; it embodies the influences of my past, and it will go somewhere; it becomes incorporated in how I envision my future role. I now better understand, not only my fears and challenges, but also my options and convictions.

The only way I can participate in Development work is as I am. I can act only from where I stand, positioned along multiple axes of power and bearing my own agenda. I must be able to participate in Development work in a way that does not deny who I am and where I come from; at the same time it cannot deny the agendas and needs of the community I work with. I must learn to operate in a less colonizing space. To implement some of the strategies addressed in this chapter will likely result in much internal struggle and confusion on my part. The process requires me to search out and resolve

some of the inconsistencies between how I envision my participation theoretically in Development and the actual reality of doing
fieldwork in a community. I look upon this process and the challenges that lie ahead for me with a mixture of excitement and fear.
Ironically, I am both excited and fearful at the possibility of being
actively engaged with the issues I have found so exhilarating. This
bizarre mixture of emotion stems from my belief that Development
issues, and how I go about participating in them, are profoundly important, both to me and to others. My fear also stems, naturally, from
a process that sees me challenging myself and my relationships to the
multiple axes of power within which I position myself and am positioned. What I require of myself is not simple, but my passion is such
that I haven't much choice but to attempt the process.

A Journey into Autobiography

A Coal Miner's Daughter

Rachel Saltmarsh

In 1995, I wrote an undergraduate dissertation. It was about pit closures in Britain, and the subsequent loss of mining culture. This research was based around the town where I was born, the mining village where I had spent my early childhood, and people who had known me all my life. I wrote this dissertation autobiographically. It was the only way that I could have done it. It was an experiment, and it began my future journey into the past, present, and future of my ailing culture.

The Beginning: October 1992

> I think that's the tragedy of working class kids having to leave school when they are just becoming interested . . . perhaps we had more potential in life. But there isn't for the majority, is there? (Kane and Kane 1994, 71)

> . . . the lower classes for whom education has traditionally been a lesson in intellectual humiliation. (Buck-Morss 1989, 35)

For our first tutorial in Lampeter we were asked to bring a newspaper clipping of an issue currently in the news.[1] At that time, the government's pit closure program had just been announced. It was

---

1. Material in the first three sections of this piece are part of my undergraduate dissertation (Saltmarsh 1995) and have been reworked into what is now this chapter.

something that everyone was talking about. So, not surprisingly, each one of us arrived with a clipping about the pits. There were four in the group—two from Yorkshire, the other two, from the south of England. The talk began, and the men started to have their say (I was the only woman in the group, apart from the tutor). Roger[2] started talking about nationalism, and about how he believed that Yorkshire should be a country in its own right, as it had all the resources (including coal) that it needed. I can't really remember what he said about the pit closure program. Gary then talked about how the pits were "uneconomic," and how it was "inevitable" that they should close. John sounded as if he had swallowed the Conservative Party Manifesto. I sat and listened, and I didn't know how to respond. I was upset. The tutor turned to me and asked me what I thought about pit closure. The words came pouring out. I said that I thought it was terrible, absolutely terrible. I couldn't explain how I felt about it. I said that mining was a way of life, a culture. Pit closure would destroy communities hundreds of years old. It was history going down the drain. Silence. Everyone looked at me—they didn't understand what I was talking about. They didn't, perhaps, understand my emotion either. I wondered if I'd said the wrong thing. I felt small. I felt, for the first time, the reality of my working-class background slap me in the face. I didn't belong here. To break the silence, I said, "Maybe it's difficult to understand mining culture if you've never experienced it." The tutor asked me if I had experienced it. "Yes," I said. "I was born in a pit village." Silence again. Then Gary said, "It's funny how this issue has brought us all together for different reasons, isn't it?" But for me, it hadn't brought us together, it had only shown our differences.

The Real Beginning: August 1973

If he was a religious man he would be a missionary; if politically minded he would be an agitator; when he likes a drink he becomes a great boozer; when he likes sports he will be a fanatical sportsman; when he is fond of gardening he is often a prizewinner. (Zweig 1948, 104)

2. These men's names are pseudonyms.

In August 1973 I was born in a pit village called Rossington. The village is one of many surrounding the mining town of Doncaster in South Yorkshire, England. My father was a miner and we lived in a pit house on King George's Road. The pit employed around three thousand people then, so the majority of men in the village worked there. I think that I was born a socialist. It is all that I have ever known. The politics of the pit village are cast in iron. This goes for the politics of homelife, as well as the politics of the workplace. The miners' militant natures and leftist politics are well known, even outside Britain. This is the environment I grew up in.

My father, whose nickname was "Townie" because he had a new suit of clothes every year, was a keen gardener. In his greenhouse he grew tomatoes and at the bottom of the garden in a strange pit, he grew leeks. He went all the way to Durham to buy his leek plants, and fed them a peculiar concoction of beer slops, manure and pigs' blood. Hardly surprising that my mother refused to eat them. Every year he would enter his tomatoes and leeks in the village prize vegetable competition. He had won first prize for tomatoes twice, and come runner-up numerous times. The best he'd ever done with leeks was to come second place. I remember my father coming

My mother, my father, and me, 1975

home from work. His eyes were bluer than the sea because of the coal dust caught in his eyelashes. He would chase my brother, Richard, and me around the garden, trying to catch us and brush our faces with his unshaven chin. We ran, laughed, and screamed, all at the same time. We had a bluey-gray whippet called Benny, my dad's pride and joy. His racing name was Salty-Blue, and he was bred out of Miss Zeus and racing champion Yorkie Boy. On Sundays my dad would take him to the Miners' Welfare playing fields, where he would race against other whippets. It was as if his little feet were touching red-hot coals on a Sunday morning. He just couldn't stand still, he knew it was race day. He was a good runner, Benny. He won stacks of trophies for us.

My mother was a housewife then. She was a seamstress by trade and worked from home. I remember her taking us on endless walks through the bluebell woods (pit or park wood as it is known locally). Nancy, our next door neighbor, and her son, Tony, would come, too. We would pick small bunches of flowers, have adventures, and take picnics with small bags of crisps. I remember my mother's baking, she would bake buns topped with white icing, each with a red glacé cherry on top, for our school parties. She would make me a different party dress each year, always a long one which would be cut down afterwards. I remember my mother, Richard, and me playing hide-and-seek in the house. She would always find us, because whenever she came close, we would start giggling uncontrollably. It was a lonely life for my mother, living in the pit village. She had never experienced mining culture before. All she had was Richard and me. My father had his mates, the betting shop, and the pub. My father loved mining culture as much as my mother hated it.

My memories of Rossington as a child are selective. I remember going to the Miners' Welfare, it looked so big. There was a huge cricket pitch, a bowling green, a football pitch, and a children's play park with swings, slides, roundabouts, and a paddling pool, too. I remember that on May Day, all the miners' children in the village got free sweets from the union. I lived in Rossington until I was eight years old. My dad, after suffering a back injury, left the pit and got a job on the railway. We moved to another part of Doncaster then.

Thirteen years on, and my father is very ill. He gasps for breath like a man under water. He has "black lung" *(pneumoconiosis)*, a miners' disease. He has a self-inflicted illness, too. Cirrhosis of the liver. He's an alcoholic. He says that he misses the pit now, he misses his mates, and he misses the club.

Educating Rachel

Until I went to university, I had never thought about my child-hood. After all, I lived in a mining town and lots of kids' dads were, or had been miners, just like mine. My friends and my family were working class, and the kids I went to school with were working class, too. When I came to university I thought for some reason it would be the same. My geography teacher told me that I would meet people like myself there, and that I'd have a really good time. I arrived at uni-versity and suffered massive culture shock. I felt like an alien from an-other planet. I didn't find many people like myself there. How could I? Working-class children are brought up to work, not to be edu-cated. My education is a result of my own desire for a different life. It was my escape route. My mother and my grandfather supported my desire. My father would not have let me become educated if he had lived at home. He thought that education was a waste of time.

My first year at university was a nightmare, and all I wanted was to wake up. Since then I've learned to live with it. It got easier as time went by. I told some people about my childhood, about my dad racing whippets and stuff like that. Their reactions made me feel as if I were some kind of curiosity. They couldn't relate to my past life; I couldn't relate to theirs. My past wasn't real to them, but how could it be? I sat in lectures and listened to people talking about labor re-lations, about inner cities, but nothing was real. People's lives were invisible. At the end of each term, I went home to these things, they were a reality to me, but to most of the people around me, they were just interesting subjects. I am glad that I have been educated, and that I have had the chance to be, but I just wish that it hadn't been such a painful process.

Deep Disillusionment, Anger, and Release

Looking back, I believe I've spent my time at university going through three stages. My first year as an undergraduate was a bad one. I went to college expecting that all lecturers must know a great deal about the world and that what they know must be right. I soon found that this was certainly not the case. I witnessed middle-class men lecturing on such subjects as class, labor relations, inner cities, and so on. This made me realize that academics often knew little about what they talk about. It wasn't just that, though; it was often the way they talked about the people or communities concerned. They showed no thought or knowledge of these people's day-to-day struggles, their lives, their tragedies. Lectures just seemed as if they were words flying at me, over me, empty, hollow, uncaring, un-feeling. This experience made me sink into deep disillusionment. I started to give up. I couldn't think what had possessed me to fight so hard for this "precious" education.

The second phase, which came at the end of my first year and lasted through most of my second year, was one of anger. I was fum-ing at academia and academics, and every piece of work I wrote was brimming with this anger. It made *me* feel better, and I really didn't give a damn who liked it and who didn't. I didn't even care whether I passed or failed. Up until this point, I think the work I had pro-duced was pretty run-of-the-mill. I didn't know how to write what I knew or whether it would be acceptable if I had. So I didn't. I just read the books like everyone else, and wrote about other academics' opinions. When I came to recognize my anger, I decided that writ-ing about others was stupid, so I wrote what I really thought. These were my first steps into autobiographical writing. My newly semi-autobiographical writings were taken very well by my tutor at that time, although I did start getting a reputation for being "too angry." The anger in my writing seemed to provoke strange reactions in some of the people who marked my work. Although they didn't seem to oppose what I was saying, it did seem to make them feel un-comfortable, perhaps because some of the things I said appeared to be a direct attack upon their position within the academy. Also, at that time, I started to get the reputation of beginning every essay I wrote with the words, "As a working-class woman, I think . . ." I never realized that I did this at the time, but I did feel that I couldn't write an essay without first telling the reader where I was coming from. After all, my early life and upbringing had had a huge impact upon the way that I thought. It seemed totally wrong to deny that.

By my third year at university, I'd realized that for my work to be taken seriously, I would have to channel my anger. Being angry was just too easy a put-down. People could say that I was angry and therefore that my work was irrelevant. With the help of my dissertation tutor (who was very encouraging and supportive of my writing), I began to learn how to express my ideas and feelings within an academic context. I went deeper into autobiographical writing and came across the concept of "situated knowledge" (Haraway 1988), which has been the foundation stone for my work ever since. What I love about situating knowledge is that it is so obvious, you write about what you know. This is completely obvious to the everyday people in the street, and yet, it causes such a fuss in a great many academic departments. From writing autobiographically, and taking on a situated knowledge standpoint, I finally feel released from the chains of academic writing. Situated knowledge was and is my rebellion too, acceptable rebellion, as long as my arguments are clear and well supported, so as not to be easily dismissed or devalued by those who don't agree with my situated approach. It is also a way for me to break with academic styles of writing and knowledge construction that I have found so oppressive, and a way that I can write my knowledge without becoming the oppressor myself. That I never want to happen.

Autobiography and Me: Past and Present

Autobiography, for me, has not only been a way of expressing my knowledge, it has also been a way for me to deal with my past and present. Going to university made me realize that I couldn't just turn my back on my past. However much I may have wanted to blend in with the majority of middle-class people there, I soon realized that my working-class culture was all that I had. The violent break from my culture and class had left me in a spin, and it is only now, looking back to my past, I can see things more clearly. It was a strange day, the day I left for college with my mother. Nearly three hundred miles away we traveled, to Wales from South Yorkshire. At the moment I was to leave her and board the bus to Lampeter, I was frantic. I cried and held her as if it were for the last time. I pulled myself together on the bus and managed to stay that way until I had finally dragged my heavy suitcase to my halls of residence and up the three flights of stairs to my room. In my new room I broke down completely and must have cried for about two or three hours whilst unpacking my things. At the time I couldn't explain why this trip had had such an impact. Later, I feel that it was probably an instinctive reaction to—as yet unrealized—violent break with everything that was familiar to me. I was upset at leaving my family, but this rather paled in significance when, on meeting people, I was hit by the reality of my class and culture. I felt as if I were treading water, a fraudster, who said nothing, but listened; who felt alienated when listening to stories from other people's pasts. This break with my class culture has been my most painful experience yet, and I could only think of it as being negative for many years. It was a loss, a hollow-feeling loss.

My present becomes my past constantly, and now I am in a different position. I began a Ph.D. on the changing nature of traditional working-class cultures (especially coal mining) in Britain over the past twenty years. I am now at the University of Wales, Aberystwyth, just down the road from Lampeter where I started. It is quite ironic, and it is an irony that makes me smile, that I am now doing this program. If someone had suggested to me a year into my first degree that I would end up doing a Ph.D., I would have claimed him or her completely mad. After everything I had experienced at university, with my lack of respect for many academics, I truly thought that I'd be the last person to be doing such a thing. Also, I never thought my work would be good enough, or that my situated-standpoint position would be acceptable. I was wrong. The situation arose, I was full of enthusiasm, and I even fought for it, as I had fought for the right to be educated in the first place. This time, I would be there, and writing on my own terms. This situation is working out pretty well, and it really is a pleasure to be writing and researching what I know and love. I am currently in the middle of my fieldwork, which involves gathering working-class people's thoughts on how their communities and cultures have changed over the years. The people I am talking to include miners and their wives, ex-miners, others who have always had working-class jobs, and their "children" (these being adult children).

Past and Present Collide: Current Dilemmas

A couple of weeks ago I went to interview Billy Phillips, a retired, Communist miner. Another Communist ex-miner gave me
his name.

"You should go and talk to Billy, he's lived in this village all his life."
I found his house with the help of other people on the bus. I walked
up the garden path with butterflies in my stomach. I knocked on the
door, a dog began barking immediately. A gruff old man answered
the door, "Who are you? Where are you from? What do you want?"
I explain to him that I'm at university, doing Ph.D. research into
mining culture. He informs me that I don't look old enough to be
doing a Ph.D., that I only look fifteen years old. I assure him that I'm
twenty-four. He calms a bit, and sits down. I then say the magic
words. "I'm doing this because my dad was a miner, and I lived in
Rossington when I was small." His attitude changes, he relaxes, it's
OK because I'm really one of them. I'm working class, but educated.
No suspicion now. He'll tell me anything I want to know. You see, we
have shared something important, we have a common bond. Our
culture. May Day, brass bands and banners parading the streets, galas,
pram races, Wellington boot throwing, pigeon racing, allotments.
Coal. (Saltmarsh, November 1997)

Since my fieldwork started, I feel that I've experienced a loss of
faith. I can't explain what this loss of faith is exactly, but it has something to do with my changing identity. Going back to the places and
people where I came from has made me realize that I have been
changing gradually over time, without even realizing it. The things I
once believed in so strongly, such as socialist ideas for a better world,
seem to have faded somewhat. This is not because I don't believe in
these principles anymore. It is more perhaps because I just can't ever
see it happening in the world today. This saddens me. Deeply. Talking to these Communist miners, who have devoted their lives to the
emancipation of the working classes, I admire their principles intensely but have had to accept that I am taking a different road from
theirs in what I am attempting to do for my fellow members of the
working class.

My fragmented and duplicitous identity has shown itself to me acutely now that I am moving between my two opposite worlds. In the conservative environment of the university, I am seen as some kind of a working-class radical. I am branded a feminist (derogatorily) by my virtually all male fellow postgraduates, simply because I can be very outspoken in my opinions. Having been brought up in the masculine environment of the pit village, I cannot help but note these middle-class men often seem to possess many of the traits that they would gladly associate with women, for example, bitchiness, back-biting, gossiping, and moaning. This is not to say that "macho" working-class men are any better, just different. I have found that each group of men can be as sexist as the other. I am constantly accused of having a problem about "class" for I'm "always going on about it." I won't be excused, nor do I ask forgiveness for bringing up such a touchy subject, one that many academics would rather not discuss. They probably wouldn't have to face up to it either, if it weren't for irritating people like me being around.

This "radical" identity sits uneasily alongside my other identity, the one that has been brought home to me recently—the academic one. The one I chose for myself, whether I like it or not. However much I may fight against it, I did choose to enter the academy and it has had its inevitable influences on me. I am a different person from the one who left the working-class, council housing estate to go to university. I'm a different person from the one who came to begin my Ph.D. a year ago. I am re-negotiating my identity constantly, just as most people are. Change is part of life itself, and if we don't change we will surely stagnate. My friend suggested to me that I should think of this change as positive for this very reason. After all, I do feel that I am giving something to the working class by collecting our histories and giving us a voice. Hopefully, too, I will be writing our stories in a way that we can all appreciate.

Writing My Culture

How could I not write my culture autobiographically? This is a strange thought. For me it would be a denial—a denial of self, a denial of past experience. For me it would be a lie. Autobiography is a way for me to share my knowledge of my culture. Through snatches of autobiography I can share its richness, its pain, its pleasures, its everydayness. My constant battle is only with those who are too blind to see its value: the scientists—those who say that my research cannot be "scientific" because I'm too involved, I'm not objective; and so on, and so on, and so on.

Here is a typical example from a few weeks ago: I mentioned to a fellow postgraduate that I had to write a chapter for a book on autobiography and geography. I said that I wasn't too worried about this, though, as my chapter would be easy and enjoyable to write. "Yes," my fellow postgraduate remarked, "it's always easier writing fiction than it is writing science." Fiction? What's fiction? What is truth? Whose science? I often have the feeling that some people are under the impression that I am writing fairy stories for kids rather than doing a piece of academic research.

I am not writing fairy stories. I am using parts of my autobiography, and the oral histories of my interviewees to write about recent cultural changes in the communities and in the places where we have been born, grown up, lived, and worked. Obsession with a thing called "science" I have come to conclude, is for those with deluded and unimaginative minds. The future of academic writing must involve situating the self, and hand in hand with this process walks autobiography. This is not to say that everyone must or should write autobiographically, but it does mean that we should be keeping our autobiographies in mind when we are researching and writing. I'm one of the few who place mine up front. For me, this is not autobiography for the sake of it, it is intentional autobiography.

Autobiography, Autoethnography, and Intersubjectivity

Analyzing Communication in Northern Pakistan

David Butz

> *Don't write this down; remember it in your head.*
> *Don't take a picture; remember this in your heart.*
> *Don't leave a message; things come apart.*
> *Talk to me face to face.*
> —Indigo Girls, "Dead Man's Hill," 1994

My purpose in this chapter is to explore how my interpretations of communicative processes in a small mountain community in Pakistan have been constituted through, and constitutive of, important facets of my recent autobiography as they relate to my interactions with the community. Specifically, I show how the efforts of some Shimshali villagers to colonize my subject position by communicating with me in certain ways—by imbricating me in certain regimes of intersubjectivity—informs an analysis of communication in their community.[1] Indeed, my own communicative experiences in

Many thanks to Kathryn Besio, Nancy Cook, and Pamela Moss for their helpful comments on an earlier draft, and to the people of Shimshal for engaging me in the terrifying and deeply gratifying process of lifeworld reconstitution.

1. During my fourteen months in Shimshal over the past decade I have had face-to-face interactions with most of the men in the community, and with perhaps a third of Shimshali women. I must stress, however, that my most intense, sustained, and complex interactions have been with two dozen or so Shimshali men, most of them leading community and lineage elders and their closest relatives (including

Shimshal keep returning me to Habermasian themes of communicative action and intersubjectivity, despite my general theoretical sympathy with alternative poststructural readings of the power-embeddedness of all communication. As a result, one of my theoretical challenges has been to conceptualize how communities like Shimshal may nurture a strong commitment to the ideal of intersubjectivity and communicative action, while retaining a clear sense of the asymmetries of power that contextualize their interactions within and outside the community. This effort reflects a recognition that my autobiography can be treated neither as a sole, trustworthy analytical tool (to the exclusion of larger theoretical issues) nor merely as an obstacle to be circumvented.

I begin with a brief overview of my research involvement in Shimshal, emphasizing my use of Habermas's concepts of communicative action and intersubjectivity to interpret recent changes to Shimshali social organization. I then describe villagers' attempts to incorporate me into a process of communicative action—in effect, to insert themselves into my autobiography—by nurturing contingencies that encourage me to (a) create a place for the community in the way I understand myself; (b) create a place for myself in the way I understand the community; and (c) involve myself in the way social and political relations unfold there. Next, I explore villagers' insistence that I endeavor to understand the community intersubjectively, and that in so doing I participate in Shimshal's political struggles with the outside world in ways that influence my own politics, career, and worldview. In particular, I examine Shimshalis' efforts—at least partially successful—to involve me directly in their

---

several women). Discussion in this chapter refers mainly to these interactions. I cannot say with confidence that other Shimshalis, were they to interact more closely with me, would deal with me in the way I describe here. I can say, however, that few of the many interactions I have had with other community members contradicted the stance and strategy described here. It is worth noting that my closest acquaintances in Shimshal are precisely those individuals who have been identified by other villagers as having the formal responsibility of dealing with foreigners like me. Thus, although the Shimshali stance described here is not representative of the community in a "sampling" sense, it is the predominant stance of those authorized to represent the community. Of course, the context within which specific individuals are authorized to deal with foreigners is not domination-free.

autoethnographic representations of the community to the outside world, and the implications of those efforts for the relationship between my autobiography and their ethnography. I conclude with a discussion of the analytical implications of my relationship with Shimshalis, emphasizing the disquieting effects such a relationship has had for my theorization of communication (see Butz 1995; Butz and Eyles 1997), and more recently resistance (see Butz 1997), in the community of Shimshal.

Communication and Social Organization in Shimshal

Shimshal is a farming and herding village of some 110 households, located in the Central Karakoram mountains in Pakistan's Northern Areas. Villagers cultivate about 150 hectares of terraced and irrigated land between 3000 and 3300 meters above sea level, almost exclusively for subsistence, and graze livestock on 2700 km^2 of territory surrounding their village. Efforts to market produce outside the community are hampered by the village's distance of two days' walk from motorized transport. Thus, the community's growing monetary economy depends mainly on nonagricultural activities. Chief among these is portering—transporting tourists' baggage for pay (see Butz 1995; MacDonald and Butz 1998). Since 1985, all Shimshali households have belonged to one of three Aga Khan Rural Support Programme (AKRSP) Village Organizations (VOs) that have formed in the community. The AKRSP's main objective has been to incorporate villages like Shimshal into a regional agricultural economy by funding infrastructural projects (in Shimshal, the construction of a link road) and by providing low-interest credit and agricultural training to communities that form a village organization that meets regularly and that satisfies a variety of other organizational conditions (e.g., paying weekly dues, attending regular meetings, consensually appointing village-level administrators, opening a village organization savings account, and collectively laboring on infrastructural projects).

My purpose, when I first went to Shimshal in 1988 and 1989, was to evaluate the implications of this new organizational structure, the AKRSP VOs, for the community's ability to sustain itself socially, politically, economically, and ecologically (see Butz 1993). In my ef-

forts to understand indigenous community sustainability in proces-
sual terms, I focused on indigenous modes of communication, and
especially communication geared to village-level decision making.
After several months of ethnographic research in Shimshal I began
to develop an understanding of these processes and their relation to
community sustainability in terms of Habermas's theory of discursive legitimacy.[2]

Although my intention here is not to rehearse the complexities
of Habermas's social theory (for this, see Butz 1993, 1995), a brief
overview of parts of his theory of communicative action provides
necessary context for understanding the relevance of autobiography
to my continuing analysis of communication in Shimshal. Stated
simply, *discursive legitimacy* arises when participants in conversation
achieve intersubjective understanding through communicative
action. Included as *communicative action* are all activities geared toward reaching understanding with other actors through a cooperative process of discussion (Thompson 1983). Thus, according to
Habermas, communicative action offers participants the potential to
overcome their specific subjective views, to assess themselves intersubjectively. Such an intersubjective assessment can be based on any
of three irreducible *validity claims:* a statement is rational if it is accepted as true in the objective sense, if it is accepted as socially correct, or if it is accepted as a sincere expression of the speaker's
subjective world (Habermas 1981, 302–9). Each validity claim relates to utterances emanating from a particular realm of experience.
Thus, we conceive of actors striving to understand some practical
situation with which they are confronted, so that they can coordi-
nate their actions consensually (White 1988, 39). They reach this
understanding by relating to objective, social, and subjective realms
of experience, through the corresponding validity claims (White
1988, 39–44). Habermas argues that this process of achieving inter-

2. In this section I am applying Habermas's very narrow definition to the terms
"discourse" and "discursive": that is, rationally motivated argumentation within a
context of free speech devoid of variations of power and linguistic competence
among participants (see Habermas 1981, 42). Strictly speaking, I am being redundant in my use of the term "communicative discourse" because Habermas's definition of "discourse" presupposes communicative action.

subjective understanding through communicative discourse is just only if it occurs within an *ideal speech situation;* that is, one without variations in power or linguistic competence among participants (Habermas 1981, 273–331). He concedes that an ideal speech situation never occurs but remains a sought-after ideal, just as oppression and domination always exist in all social relationships.

Habermas's model is especially relevant for my interest in the communication processes by which indigenous communities sustain themselves because of his assertion that justice and rationality in decision-making rest on the procedures through which decisions are made (Habermas 1981, 1987); the justice of roles and outcomes is contingent upon the rationality of a decision-making process. For Habermas, the process of communicative action—intersubjective validation of one another's experiences—enables the rationalized reproduction of a matrix of background convictions, described as a shared lifeworld, and manifest more concretely as generalizable interests: things that participants in discourse agree are desirable. These interests, intersubjectively shared by participants in communicative discourse, become the primary underlying influences on specific decisions. But, generalizable interests, and the lifeworld they represent, are not static. They are revised and reproduced each time participants attempt to use them as bases for validating utterances. An important implication is that as we cannot expect different speech communities (or the same ones, at different times) to share identical generalizable interests, neither can we expect what they call just or rational decisions to be identical.

I have always been uncomfortable with applying as modernist and Eurocentric a metatheory as Habermas's to an analysis of social organization in a small village in northern Pakistan. Had I felt competent to base my analysis solely on "distanced" observations of how Shimshalis communicate with one any other, I think I would have abandoned a Habermasian interpretation long ago, despite the community's obvious commitment to village-level consensual decision making. What prevents me from doing so is my own experience of Shimshalis' sustained communicative efforts, over a period of a decade, to reconstitute my lifeworld—my autobiography—and theirs, so that something approaching intersubjective communication is possible between us. As my research interests in Shimshal have shifted from a general focus on community sustainability to a more specific concentration on Shimshali resistance to domination within the adventure tourism industry this experience has intensified, and I recognize that my experience of Shimshalis' attempts to create conditions conducive to communicative action with outsiders is not unique (see Butz 1997). It is a tactic common to villagers' interactions with Western tourists, which confirms my sense of Shimshalis' general commitment to an ideal of intersubjective communication in their relations with other Shimshalis and with outsiders. In the following section, I enumerate some of the ways that Shimshalis have inserted themselves, their community and their lifeworld into my autobiography.

The Colonization of My Autobiography

Shortly after I began research in Shimshal—at around the time several leading community members decided I was unlikely to be a danger, and could become a resource—the villagers I had closest contact with began a process of taking over my research project.[3]

3. My use of the term "colonization" to describe the ways that Shimshalis' discourses have infiltrated my subject position requires some explanation, especially as I enjoy obviously greater access than do Shimshalis to the material and discursive resources of power in a global field of domination. I think it is appropriate—rather than offensive—to speak of the "colonization of my autobiography" for several reasons.

First, I am attempting a fairly straightforward play on Habermas's notion of the "inner colonization of the lifeworld." Habermas uses the concept to describe how the process of lifeworld disenchantment necessary for the nurturance of communicative action is, in a capitalist world system, undermined as "the media of money and power increasingly infiltrate spheres of social life in which traditions and knowledge are transferred, in which normative bonds are intersubjectively established, and in which responsible persons are formed" (White 1988, 112). As such, the inner colonization of the lifeworld is characterized as a unilateral process—in global social and spatial terms, from the "West" to the "East"—akin to colonialism in general. Thus, both the potential for communicative action and its most powerful frustration are conceptualized as originating in the modernized "West." My use of colonization acknowledges the transcultural nature of lifeworld colonization; that conditions sometimes emerge for subaltern populations to use their own form

This occurred initially through the straightforward procedure of regulating my access to information and experiences, but gradually developed into a more subtle tactic of encouraging—even requiring—me to situate myself within a specific set of discursive relationships with the community.[4] With the passage of time the latter tactic

---

of inner colonization to resist the lifeworld-colonizing influence of the systemic media of money and bureaucratic power, in this case (although not necessarily) to increase the potential for intersubjective understanding.

Second, the implications of this play on inner colonization are commensurate with Bhabha's assertion that colonial power is never possessed entirely by the colonizer because of an ambivalence that lies at the root of the West's approach to subaltern Otherness, which is characterized by a will to produce "the colonized as a fixed reality which is at once an 'other' and yet entirely knowable and visible" (Bhabha 1983, 199). Bhabha suggests that colonial power produces hybridized subjectivities, through which "other 'denied' knowledges enter upon the dominant discourse and estrange the basis of its authority" (Bhabha 1985, 156). Hybridity can thus be used by the colonized to become a "strategic reversal of the process of domination" (Bhabha 1985, 154); it transforms colonial discourse into something that disrupts what colonizers intended, by tactically inserting repressed knowledges into colonial discourses. What I am describing as the colonization of my autobiography is just such a tactical insertion, and I experience it that way.

Third, and stemming from the previous two points, is the necessity to recognize colonization as a process that colonizes members of both the "colonized" and the "colonizing" societies. The latter is aided by the tactical maneuvers of the colonized, which are denied by too unilateral an understanding of colonization. And it is important to assert that this bilateral process occurs locally despite radical global asymmetries in access to material and discursive resources, and without denying the massive implications of these asymmetries, both locally and globally. In sum, western researchers working in situations similar to that described in this chapter need to recognize (and embrace) their own vulnerability to colonization by their "subjects" as a first step in efforts to develop non-Othering interpretations.

4. Here, and for the remainder of the chapter, I am following Gregory's description of discourse as "all the ways in which we communicate with one another, to that vast network of signs, symbols and practices through which we make our world(s) meaningful to ourselves and to others" (Gregory 1994, 11). According to Foucault (1972, 46; emphasis in original) "whenever between objects, types of statement, concepts, or thematic choices, one can define a regularity . . . we are dealing with a *discursive formation.*" Discourses and discursive formations are constitutive of the conduct of day-to-day life (Barnes and Duncan 1992, 8), central to the naturalization of particular worldviews (Foucault 1967), and situated in specific fields of power and knowledge that may be contested and negotiated. They are also shifting, contingent, and partial. What Habermas calls discourse—argumentation

has intensified to the extent that I find myself an active participant in this incorporation, while the former tactic of regulating access to instrumental information has diminished significantly. My relationship with the community's fledgling environmental education program exemplifies this gradual shift. I spent a disproportionate amount of my first seasons in Shimshal attempting to acquire background statistical information on population, household composition, agricultural productivity, and so forth. My efforts were obstructed at every turn, sometimes without explanation, and sometimes with the justification that I could not be trusted with such information. More often Shimshalis argued that the information I sought was simply not relevant for understanding the most important things about Shimshal; a point I thought was well taken. Meanwhile, in a different part of the village, school children were busy collecting—under the auspices of the "Shimshal Environmental Education Program"—exactly the type of information I thought I needed. For a long time I remained unaware of these data-gathering activities. Later, I was told of students' activities but denied access to their results. Two summers ago I was finally given full access to all data they had collected. During my most recent visit to Shimshal I found myself in the unexpected and conflictual position of revising my research interests—at the request of community elders—for the sole purpose of providing students with a reason to expand their data-gathering efforts, *and* encouraging teachers to reconsider this recent emphasis on teaching students to understand their home in statistical terms. In this instance, as in others, persuasion/incorporation have largely replaced coercion/violence as mechanisms for contextualizing my activities in Shimshal; a shift that indicates Shimshalis' and my mutual efforts to establish some shared generalizable interests, some potential bonds of intersubjectivity.

The process through which villagers have strategically situated

------

within an ideal speech situation —is a very specific discursive practice, and one way that discourses (in the Foucauldian sense) are constructed and manipulated. Part of my argument is that Shimshalis' attempts to engage me in communicative discourse have both inserted my autobiography into local discursive formations, and aided in what I am describing as the colonization of my autobiography by these same discursive configurations.

me in a common discourse (narrative, ethnography, autobiography) with the community of Shimshal has had three main elements. First, they have insisted that I create a space for the community in the way I understand myself. This has meant incorporating me, and recently my family, into the household and lineage activities, concerns, and obligations of the people I am closest to while I am in the community. By investing me with household-based obligations in the community's public sphere (such as participating in and contributing to agricultural festivals), and by involving me in discussions about the future of junior members of "my" household, my research participants enlarged the scope of my lifeworld to include their household-level and community-level generalizable interests, and created an opportunity—some common ground—to engage me in communicative action. This can be explained, in part, in terms of a local ethic of hospitality to outsiders. But, the shared experience of giving/receiving hospitality is also a basis for intersubjectivity, a basis for establishing communicative connections. And discourses of hospitality are closely related to discourses of obligation. For example, at a certain point it became clear that part of the reciprocal relationship of hospitality I enjoyed with the household in which I lived was an obligation to participate in "finishing" the household's eldest son: to help him prepare to assume his father's position as an influential (and highly educated) lineage elder. Because the household's position in the village was prominent, this obligation was also expressed—and experienced—at the scales of lineage and community. We decided that the young man would serve as my (well-paid) research assistant; a decision that had everything to do with his education and anticipated social position, and little to do with my research requirements (which would have benefited from a more scholarly assistant). Our research association did help "finish" the young man, as well as provide the context for us to become friends, as his parents anticipated. It also effectively created a space for Shimshal in the way I understand myself: my research interests and a household's aspirations have become closely entwined; the web of responsibilities in which my identity is enmeshed has grown; and I now understand one of my power-effects to be the way my patronage of this young man has influenced power relations in the village of Shimshal.

Shimshalis' efforts to insert the life of their community into my own lifeworld extend beyond the spatial borders of Shimshal. Villagers also make a point of comprising part of my everyday life when I am elsewhere in Pakistan, by visiting me, by seeking my assistance in their down-country activities, and by occasionally using my town home as the hub of their own lives outside the village. Although I have not yet entertained Shimshali visitors in Canada, I do receive frequent news from Shimshal, and offers and expectations of advice and assistance by mail. As I write this from my office in southern Ontario, I am in the midst of trying to negotiate a way for one of Shimshal's most educated young men to study "enterprise development" at a Canadian university. In this way, and others, my lifeworld, my life here in Canada, and Shimshalis' dreams, plans, and social strategies are mutually imbricated, in a manner that is highly time/space distanciated. Again, the experience of offering/expecting hospitality is a source for the development of further intersubjectivity.

These concrete, and unrelenting, activities have the effect of reconstituting my sense of self to incorporate my relations with the community. Shimshalis have also taken pains to connect my life and theirs at a more structural and historical level, by insisting that I recognize the relationship between my position within a history of colonizing and theirs within a history of colonization. My experience with this effort to situate historically their circumstances within my life is shared by the many trekkers to Shimshal who are reminded of the colonial lineage of contemporary portering labor relations, and who are sometimes asked quite explicitly to choose among identifying with archetypal colonial era explorers, down-country tour organizers, or the Shimshalis who are carrying their loads and offering them hospitality (see Butz 1997). For example, porters who speak some English often pass time on the trail by telling trekkers—and other travelers, like me, who walk along the trails to Shimshal—stories of Shimshal's history; stories that hint at an "authenticity" that is contrasted to the hybridized present;[5] stories that introduce us to knowledge whose origins predate the field of colonial domination.

---

5. "Authenticity" is placed in quotation marks to indicate that although travelers frequently search for an "authentic" experience, the concept of "native authenticity" is itself problematic.

We are thus initiated into a vision of potential authenticity, forcibly denied by outsiders like ourselves. We are also told in some detail of the ways threats to this authenticity have been resisted, and of the precedents for everyday resistance in indigenous history. The candor with which porters reveal this gives us (trekkers and researchers alike) a taste of the "authenticity" we may desire. Later, when we are victimized by the same forms of everyday resistance, we may be thrilled—or ashamed—to recognize and experience subaltern "authenticity" more fully. I understand this as an attempt to exert some control over the discursive representation of everyday practices of resistance, to clearly identify these practices as resistance and not oriental dissolution, and to provide trekkers and other visitors with a sense of our own complicity in the reproduction of a set of oppressive labor relations. Of course, all these activities can be read as strategic efforts to gain some instrumental resources from the most proximate representative of domination. An important part of my argument is that Shimshalis' hopes of achieving any of these instrumental gains relies on a commensurate hope of establishing, despite great differences in subject positioning, some common understandings of certain aspects of the world sufficient to form the basis of consensual decision making.

A second, and only heuristically separable, way that Shimshalis have situated me in a common discourse with their community is by encouraging me to create a space for myself in the way I understand Shimshal. In the same way Shimshalis remind me that their history of subordination is connected to my history of domination, they also remind me that my program of research influences their processes of communication, and more recently, their efforts to resist domination. Moreover, community members have quite deliberately involved me in ways that will ensure that my presence in the village reconstitutes dominant community-level discourses, by inviting me to argue in the public sphere in Shimshal, and to represent Shimshal in certain venues outside the community. In a related move, villagers have not hesitated to variously credit me with, and blame me for, discursive shifts in Shimshal and with the outside world. I experience this as exposure to continual and explicit evaluation; a sort of "let's talk about how you're doing, and what you need to work on" that enumerates the trajectory of my status in various spheres of public opinion in Shimshal. My examiners' emphasis throughout has been on the assertion, sometimes the accusation, that Shimshal is our common project, mine as well as theirs. This deliberately crafted commonality of purpose is another element that unites my lifeworld with Shimshalis' and prepares the way for potential communicative action.

Third, these mainly dialogic efforts to spin threads of mutual obligation and subject/community constitution between Shimshal and me have their concrete manifestations materially and institutionally. The fact that I have accepted certain material obligations to the households I have lived with to their material benefit, attempted to sway public discussions in Shimshal sometimes as a spokesperson for particular factions within the community, and acted to represent my interpretations of Shimshali interests to various agencies outside of the community, all help to establish the basis for asserting that Shimshal is indeed our common project, and an integral part of my autobiography.[6]

After a decade of exposing myself to these processes of incorporation, I recognize that my autobiography has become a jointly produced, shared resource in our struggle for intersubjective understanding, as well as an important—if somewhat ephemeral—analytical tool. My autobiography bleeds into their autoethnography.

Autobiography Bleeding into Autoethnography

Shimshalis' attempts to incorporate my autobiography into their project is an intense example of a more general autoethnographic

6. It is worth noting that each of the three ways Shimshalis have strategically situated me in a common discourse with the community is strengthened and complicated by community members' separate—but not independent—relationships with my partner (who has spent five months in Shimshal) and our daughter (who visited Shimshal briefly when she was three, and again for several months when she was five). The web of associations, obligations, and opportunities for communication that connect my lifeworld with Shimshalis' is intensified by the fact that I share Nancy and Nina with them, they share Nancy and me with Nina, Nancy shares me and them with Nina, and so forth; and this is multiplied by our associations with a range of diverse Shimshali "thems." To varying degrees we have all become one anothers' projects.

agenda to establish a strategic intersubjectivity with certain categories of outsiders. Mary Louise Pratt (1992, 7; emphasis in original) uses the term *"autoethnography"* to "refer to instances in which colonized subjects undertake to represent themselves in ways that *engage with* the colonizer's own terms." She goes on to explain that "if ethnographic texts are a means by which Europeans represent to themselves their (usually subjugated) others, autoethnographic texts are those the others construct in response to or in dialogue with those metropolitan representations" (Pratt 1992, 7). Pratt employs autoethnography quite specifically, to describe the translation of indigenous cultural texts into the language and idiom of metropolitan literate culture. I think the concept has greater heuristic utility if it is expanded to include any attempts by subordinate groups to establish a communicative basis for explaining themselves to—and to establish a common understanding of their group circumstances with—those dominant outsiders who have conventionally assumed the prerogative to represent and explain the subordinate group in their stead. The latter definition retains the situatedness of autoethnographic representations within a field of asymmetrical power relations but foregrounds subordinate groups' occasional reliance on the ideal of intersubjectivity as a tactic for resisting subordination. Autoethnographic expressions are likely to be aimed at quite specific audiences; those from which subordinate groups have some reason to expect a sincere effort at communication. To the extent that subordinate groups realize such audiences are rare, autoethnographic expression—and the search for intersubjective understanding more generally—is likely to be only a small part of a larger repertoire of everyday resistance to domination (Scott 1990).

The community of Shimshal has involved me in their autoethnographic project in two ways. First, their efforts to establish a basis for intersubjective understanding through the three processes described in the previous section are themselves autoethnographic endeavors. I have become an object (subject) of their autoethnographic expressions. The insistence of many Shimshalis that it is possible for us to engage in communicative action, despite my frequently expressed doubts, and their continual labor to engage and incorporate me into a set of Shimshali generalizable interests, are both indicative that I have been identified as an audience from whom Shimshalis can expect a relatively sincere effort at communication.

Second, Shimshalis have enlisted me as an ingredient and agent of their autoethnography. I am presently part of the story Shimshalis tell of themselves to the outside world. I am presented as an artifact—a living proof—of the community's claims to hospitality, modernity, openness to scientific research, and potential for productive and nonexploitative relations with the outside world. My story has become part of their autoethnography; part of their attempts to establish a basis for meaningful communication with others.

I have also become an agent for the refinement and circulation of their autoethnography. Shimshalis have sought my assistance in helping them craft their autoethnography into a language and idiom that might allow them to communicate with those that dominate them. In some cases this has been the language of political entitlement; in others, the language of cultural and ecological sustainability. They have also enlisted me to speak their autoethnography in places Shimshalis cannot reach, through scholarly publications and conference presentations, and also more informally in conversation with whomever I think constitutes a relevant audience. I find myself in an awkward position as coauthor of some of Shimshal's autoethnographic expressions; an indication both of the extent to which Shimshalis trust (are forced to rely on) whatever level of intersubjective understanding we have achieved, and of the vulnerability of autoethnographic expression to appropriation by members of dominant groups. The most significant example of my coauthorship in Shimshal's autoethnographies has been my involvement in the Shimshal Nature Trust. In the early summer of 1997 I returned to the village after an absence of two years. I was met by a delegation who immediately asked me to help them frame a set of formal guidelines for the internal and autonomous stewardship of their territory (in opposition to a government/NGO initiative to incorporate most of Shimshal territory into a limited-use national park). Over the course of the summer these guidelines expanded into the Shimshal Nature Trust (SAT). I was assigned the limited task of collating community members' various ambitions for the trust into a draft document, in English, that would describe and justify the trust to non-Shimshalis, as well as define a set of operating principles and procedures for villagers (I was also asked to put the document on the World Wide Web, and distribute hard copies at my discretion).[7] This was not an unproblematic act of translation, nor did the Shimshalis who solicited my input imagine it to be. We all realized—with trepidation all around—that opportunities for misrepresentation abounded. My involvement in the trust—an important autoethnographic expression—was a leap of faith: Shimshalis' desperate faith in my trustworthiness; my own uncertain faith in my fidelity to the community's (by no means unified) sense of itself; and, especially, a mutual faith in those fragments of intersubjectivity linking my autobiography to their autoethnography.

Three points emerge from this discussion of my incorporation into Shimshal's autoethnography. First, I do not think the process by which I have slipped from being a privileged audience of Shimshal's autoethnography to being a coauthor is unique to my experience. It is fairly typical of the autoethnographic expressions of subordinate groups that they are not just attempts at autorepresentation, but also aim to incorporate the intended audience into the autoethnographic project. In these cases subordinate groups' relative lack of power—the absence of an ideal speech situation—may actually force them to rely on the ideal of communicative action in order to give voice to their autoethnographic expressions, with the attendant risk (almost inevitability) that a cynical audience will appropriate and recirculate a distorted version to serve their own purposes of domination.

Second, to the extent that autoethnography is, by definition, a political strategy, my incorporation into Shimshalis' autoethnographic project inserts me into Shimshal's political struggles in ways that influence my own program of research, career, and politics more generally. Given the circumstances described throughout this chapter, I have found it impossible (and undesirable) to avoid identifying myself publicly with Shimshal's struggles, and to some extent against the development agencies, international ecological organizations, tour operators, and Pakistani administrative units with which Shimshal is struggling. This identification has influenced the

7. The Shimshal Nature Trust can be located at http://www.BrockU.CA/geography/people/dbutz/shimshal.html.

tone and content of my scholarly publications, the details of my vitae, the objectives of my research applications, the content of my teaching; in short, all public and professional expressions of my recent autobiography. This has implications, no doubt, for the long-term development of my career.

Third, and most relevant to my main argument, my various points of insertion into Shimshalis' autoethnography have shaped my own autobiography and given me an opportunity to use that autobiography in ways that profoundly influence my analysis of communication in Shimshal. In the chapter's final section, I focus on the interpretive implications of using this discursive relationship with Shimshalis as an important source of data for understanding processes of communication in the village.

Autobiography, Autoanalysis, and the Politics of Hope

I have argued throughout this chapter that many of the Shimshalis I am closest to have engaged in a lengthy and subtle, but quite overt, endeavor to colonize my autobiography by tinkering with my lifeworld. Their stated objective has been to create a set of shared circumstances that allow us to understand one another better; what Habermas would describe as establishing the lifeworld basis for communicative action. This is obviously a strategic political undertaking. Shimshalis are not expending this level of effort on me simply for the pleasure of communication, although in a cultural setting as conversationally oriented as Shimshal it would be unwise to underestimate the joy of aesthetically satisfying verbal communication. Rather, they *hope* that by crafting the basis for intersubjective understanding they will create in me a trustworthy and reliable ally in their autoethnographic, and other political, struggles. I doubt that a "distanced" examination of communication in Shimshal, were such a thing possible, would have revealed that Shimshalis' commitment to the ideal of intersubjectivity is a *politics of hope* fueled by the desperation of the powerless.

An autobiographically driven sense of this politics of hope allows me to reevaluate the potential and limitations of Habermas's theory of communicative action in situations such as these. Habermas imagines legitimate communicative action as emerging only in situations where all participants share similar characteristics of communicative competence and power. As a result, his theory of communicative action is poorly equipped to deal effectively with a poststructural conceptualization of the power-embeddedness of discourse (Poster 1989). What my experience of Shimshalis' efforts to colonize my autobiography suggests, is that it may be productive to imagine communicative action less as the unattainable outcome of an unattainable ideal speech situation, and more as a principle, and mechanism, for selectively and strategically reducing power asymmetries. For Shimshalis, commitment to the ideal of communicative action does not betray a naïveté toward their relative lack of *power* vis-à-vis the outsiders they encounter, including me. Rather, this commitment is a strategy, selectively deployed, both to neutralize somewhat the power of the outsider, and to incorporate aspects of that power into their own autoethnographic endeavors. In other words, Shimshalis engage in communicative endeavors with outsiders, not because they fail to recognize their relative lack of discursive power, but precisely because their strong sense of powerlessness in the instrumental realm causes them to recognize that their only hope of neutralizing the power of the Other is to incorporate selective Others into an intersubjective recognition of, and commitment to, Shimshalis' validity claims. Moreover, to the extent that residents of communities like Shimshal recognize that this is a faint hope indeed, they will also engage in other, less intersubjectively oriented strategies to practice some power over powerful outsiders. Contrary to what Habermas's conceptualization might suggest, Shimshalis are playing several games at once, only one of them communicative. This helps to account for my frequent sense of betrayal by Shimshalis; even as they are engaged in sincere, if strategic, efforts at communicative action with me, they are also keeping other paths open. I am too valuable as an immediate instrumental resource to be approached only from a longer-term communicative position. Occasional betrayal of the latter in the interests of the former is a strategy, the risks of which are calculated according to Shimshalis' hopes that I share their understanding of the need to play several games at once. The politics of hope implicit in the search for intersubjectivity is merely one in a variable set of practices of everyday resistance (Scott 1990). For the powerless, hope is itself a practice of resistance.

The risks implicit in playing this game of intersubjectivity with the more powerful also help us to understand the much advertised ambivalence of subaltern discourse (Bhabha 1984, 1985). In the attempt to create some basis for intersubjectivity with me Shimshalis are engaged in what is well described as a limited and strategic mimicry of my discourse in an attempt to incorporate me into theirs. Autoethnography is characteristically such a form of mimicry which hopes to reach out to an intersubjective engagement with the Other.

It is worth noting that these autobiographical reflections on the characteristics of my interactions with Shimshalis have sensitized me to similar characteristics of communication and resistance *within* the community. Specifically, I have become more aware of both the importance of the ideal of communicative action to community-level discourses, and the stark—if somewhat veiled—internal asymmetries of power that contextualize the deployment of this ideal.

I am led to conclude that my experience in Shimshal demonstrates a more general situation in which the interpretation of field research and the process of reflecting on autobiography cannot be, nor should be, separated, methodologically or epistemologically. Moreover, the theories we develop to explain our research findings are necessarily autoethnographic: attempts to reach out to incorporate others into our autobiographies, based on a politics of (faint) hope that intersubjectivity is possible despite radical power asymmetries within the academy and without.

# Many Roads

The Personal and Professional Lives of Women Geographers

Janice Monk

Early in my career as a geographer, I learned that professional and personal worlds were intertwined, and that my opportunities would reflect not only my own aspirations but also structural conditions and the expectations and needs of others around me. I grew up in Australia in the 1950s, a time when few professions other than school teaching were open for women, and when daughters were expected to live "at home" until marriage unless they were sent to a rural school to teach or ventured "overseas" for extended working holidays (Pesman 1996). Since my family was of modest means, the only road open to me to obtain a university education was via a scholarship from the New South Wales Department of Education; it paid my fees and a small living allowance, but required me to sign a bond to teach for the state for several years after graduation.

As chance would have it, at the time of my graduation, the head of the Department of Geography at the University of Sydney was looking for a research and editorial assistant to complete his languishing book project. He offered me the position, and I took a year's leave from the Department of Education, postponing enrollment for the postgraduate teaching diploma. I enjoyed the work and, at year's end, when the project was still incomplete, took up his

I appreciate the cooperation and candor of the women whose stories are included in this chapter.

offer to stay at the university in a junior position that combined research assistance with some undergraduate lab teaching. In order to do this, I resigned from the Department of Education and paid off my bond. I remained two more years, then left for graduate school in the United States.

I learned several things during this period in my early twenties, the first being that the state valued marriage and motherhood. Had I married, the bond would have been reduced significantly, had I become a mother, it would have been waived. But teaching at the university (where many of my evening students were elementary school teachers employed by the state) was not seen as having equivalent value, so I was required to pay the full amount. My professional value was linked to my personal status. The work setting also helped me begin to recognize my own aspirations and values, though still not to articulate them in long-range terms. I knew there had to be more to my future than settling in suburbia (and washing a man's socks which for some reason was my phobia). Being at the University, I saw advertisements for graduate assistantships in the United States, so I applied to several U.S. departments of geography, planning to do an M.A., then possibly to join a high school friend for the conventional Australian girls' extended working trip in England and Europe.

I was responding not only to professional opportunities and goals, however, but resisting familial constraints on my independence. Since I had bypassed the rural teaching that awaited most new graduating teachers, I was still living with my parents and I found it confining. The only other conventionally acceptable route to departure, besides marriage, was to leave the country, presumably temporarily. My father, who had a lifetime interest in travel but lacked the resources to do so, was supportive; my mother was less convinced. As it happened, a couple of months before I was aiming to leave, my father died. As a new widow, my mother had a change of perspective; she thought that I should do what my father had supported. I departed at age twenty-four, expecting to be away a couple of years, without a clear sense of my future. Since then, I have returned "home" only on short visits, for research, conferences, and to see family and old friends, though I retain my Australian citizenship and my nationality is part of my identity. But an important part of my life has been creating another "family" in which feminist geographers from an array of countries are important members.

I earned my doctorate as the women's movement was gaining force. Given my personal history and aspirations, I was perhaps a "natural" to gravitate to feminist scholarship, to see that my own story was not idiosyncratic, and to become interested in questions of the intersections of gender, cohort, life course, context, and ways that women geographers navigate the personal and the professional. Those interests have stimulated this chapter.

This chapter portrays a world that is virtually invisible in writing about geographers—the personal relationships in marriage and the family that are intertwined with career paths. I draw on interviews with six contemporary women who were in their forties at the time that they reflected on their lives. All have earned doctorates in geography in the United States and have worked professionally. Their personal lives emerge as key aspects of their careers in ways that may not superficially be as critical in the careers of men geographers, though there are hints in the literature that men's work too has been linked with their personal relationships—not least in that their wives made it possible for them to focus on their work, but also in that their wives have been field partners, unpaid research assistants, editors and bibliographers, hostesses supporting career advancement, and even political fence-menders in "geography wives'" groups.

In auto/biographical accounts of the life paths of men geographers, accounts of personal lives mostly dwell on childhood, parents or ancestors (e.g., Buttimer 1983c; Buttimer and Hägerstrand 1988).[1] If wives and children are mentioned, they appear briefly—as travel companions (Humlum 1988; Tuominin 1988), as bringing about a change from nights in the office, clubs, or restaurants to nights at home (Bergsten 1988), or occasionally as supportive coworkers (Somme 1988; Tuominin 1988).[2] Sometimes it is the ca-

1. Martin Kenzer's (1998) current work on the significance of Carl Sauer's relationship with his wife is an unusual exploration.

2. Such references are most often found in the acknowledgments sections of books.

maraderie of the bachelor state and the masculinity of graduate student culture that attract comment (Hart 1979; Morrill 1984). The sparse auto/biographical literature on women geographers also reveals little of the personal lives of the (mostly single) women who have been memorialized in brief obituaries[3] or the subjects of the few longer studies (Andrews 1989; Berman 1980, 1984; Bushong 1984; Horst 1981; Sanderson 1974). The short autobiographical accounts by British women geographers (Women and Geography Study Group 1997) also emphasize their lives in the academy. Hints of the importance of the private world in shaping the professional experience, however, emerge in Bingham's turning to literary studies (Berman 1984); in the support for Somerville's work by her uncle and husband (Sanderson 1974); and in Brun-Tshudi's appreciation of the stimulus she derived from her husband's curiosity (Buttimer 1983b). By contrast with these works in geography, the extensive literature on women in science (e.g., Abir-Am and Outram 1987; Gornick 1983; Kass-Simon and Farnes 1990; Rossiter 1995) makes clear the salience of personal life in women's professional experiences.

Perspectives on Connecting the Personal and Professional

Recent theoretical and methodological literature in geography, especially in feminist geography, has explored subjectivity in research, especially themes of positionality, reflexivity, and textual representation (for example, Nast 1994; Katz 1994; Kobayashi 1994; England 1994; Gilbert 1994; Staeheli and Lawson 1994). Gill Valentine's (1998) account of being harassed goes into depth on the reciprocal effects of this episode on the construction of both her identity and her research. I write cognizant of the issues raised in these works, and also of literature on auto/biographical writing, especially of the perspectives that recognize that all stories are partial; that "for a number of reasons we do not attempt to make all things apparent to all people" (Nast 1994, 60); that "the rightness of any au-

3. The *Journal of Geography,* especially in the 1960s, contains a number of women's obituaries, for example, of Alison Aitchison, Alice Foster, Edith Putnam Parker, and Zoe Thralls.

tobiographical version is relative to the intentions and conventions that govern its construction or its interpretation" (Bruner 1993: 40); and that gender shapes the representation of the self (see, e.g., Friedman 1988; Okely 1992; Stanley 1992). Especially germane for my work is the perspective of seeing the self-in-relation, rather than as an autonomous, isolated being, a perspective several authors associate with women's autobiographies.

Scholarship on the life course (see Katz and Monk 1993) also provides a key to my chapter. It emphasizes the diversity of experiences within an age group, exploring pathways through the various structures in the major role domains of life, and pays particular attention to points of transition, while setting the life course within historical contexts. Important analytical concepts are the significance of cohort, of prior experience for later life, of continuing development, and of the conjunction and synchronicity of diverse roles as they shape the capacity for choice and constrain options.

Interviewing Women Geographers

I interviewed the women included in this chapter as part of a project on the history of women geographers in the United States since the late nineteenth century. The project included archival work and interviews with approximately sixty women, most of whom hold doctorates in geography, and who were aged between their late thirties and early nineties at the time I talked with them. They represent different cohorts, substantive interests, and positions in the field. Because the interviews include personal matters, there was an agreement of confidentiality so that names and places are changed in my text.[4] The women's discourses varied, especially by generation, as did their feminist consciousness, detail in recounting their lives, and the extent to which they responded to my questions or designed their own narrative.

In this chapter I have selected six women from my collection of interviews—women who have potentially experienced similar

4. Individuals are named when the research uses archival sources (Monk 1998), a practice that often grants differential privilege to the written, not the spoken record.

conditions of education, employment, and disciplinary culture. The full collection includes women of color, foreign-born, and lesbian women, but those presented here are all white, though of different ethnic and class backgrounds. I have also chosen women who are or have been married and raised children, especially because I am interested in the intersections of work and family life. Had I focused on earlier cohorts, a substantial number of the women would have been single or married and childless. This is not to imply that single and childless women lack personal lives or that married women with children do not have other ties, simply that the relationships under consideration would have been different.

The stories follow a conventional chronology, beginning with childhood and educational background, then moving to marriage and family as these intersect with higher educational and career experiences, but the individual narratives make clear that the differences in timing over the women's life courses have shaped their choices and current circumstances. The importance of their personal values and those of their family members are also revealed as affecting the ways in which some pathways open and others lead to professional detours (or perhaps closures). I rely heavily on the women's own words, recognizing that these were spoken in an interview context.[5]

Growing Up

These women were born in the 1940s; several were the first generation in their families, or at least the first generation of women in their families, to complete a college education and pursue a professional career. Ellen Knox, now a mid-career faculty member in a re-

---

5. All but one of the interviews were conducted in private settings. In one case, the woman's husband was in earshot, though more or less occupied with his own activities. He did, however, intermittently interrupt, usually to provide supplemental information, reinforcing my awareness that our memories are all fragmentary, but also leaving me to wonder in what ways the story might have been told differently in his absence.

search university, is such a woman. She grew up in one of the poorer regions of the United States in modest circumstances:

> My mother was a high-school graduate, which was quite an accomplishment for her. My father had expected only to get high-school education, but he went off into the [service] in World War II, did very well on a standardized exam, and was put through college. . . . It opened up a whole new world for him . . . but he feels . . . that he wasn't prepared for the life he ended up leading. With hindsight, I think that whatever made me however independent I am is two particular great aunts, both of whom were strongly disliked by my father, both for what they were and for their influence on me. One was . . . extremely outspoken. She had helped raise her nephews, never owned anything, worked at times to help support the family, but never had anything like a career. But she was very sure of herself, very cranky, and felt that she had a right to be. And she and I were very close as I was growing up. And when my father and I would have disagreements . . . she stuck up for me, and always made me feel that I was right. The other was an aunt of my mother. . . . My father disapproved of her because she smoked, and she swore, and she read mystery stories, and married a couple of extra times. . . . She was my heroine. . . . She just encouraged me to do what I wanted to do. . . . In some ways my parents were supportive, but they didn't see why I would need to go on to college.

Ellen graduated as high-school valedictorian without knowing what she wanted to do "or what college was all about." She enrolled at a state university with no specific goals, graduating at age nineteen. Discussing her early years, she presents herself as "deviant," as independent, as eager to explore diverse fields and the world, and as "naïve about lots of things." Her family upbringing gave her some direction, however: "I'd been raised in a fundamentalist family and I thought that if I were going to do what I wanted to do, at least it should benefit people." Reflecting the period in which she graduated (and the availability of federal support), she obtained a fellowship to study community development, choosing a distant school to increase her knowledge of the world. "I got there and quickly realized I wasn't going to be able to live off my stipend. I had assumed I would be able to do that. I was incredibly naïve. But also very resilient."

Pam Linsky's childhood also gave her a sense that she was different:

> The first five years of my life, my parents lived with my great-grandmother who owned a boarding house . . . in the inner city. I grew up with all different kinds of people . . . all different ages. And then we moved to white suburbia, and there I lived for ten years. . . . When I was fifteen, my dad died and our family split up. My mom remarried in a year and I went to live with my grandmother . . . back in the inner city. . . . I think it had an effect on me. I felt different when I moved. I felt like I didn't fit in. And so I think my identification with people who think that they don't fit in, like minorities, I think I understand.

Today, issues of ethnic diversity feature in Pam's teaching and writing. Unlike Ellen, Pam went to college for only one year, married at nineteen, and became a mother at twenty. She returned to school in her late twenties.

Anna Kasic's background provided her with little experience of college ways, but direction came from a geography teacher. Anna grew up in a closely knit ethnic community and didn't learn English until she started school.

> I grew up in a home with my grandparents and my mother. . . . My mother and father had split at a time when divorce was unheard of in [our] community. We grew up in a very, very disciplined and very sheltered home, and the idea was that education was really important. I was fascinated by the world of places and I had the kind of teachers who promoted that. . . . At one point I really wanted to go into medicine, but my mother discouraged me, saying that there were very few women in medicine at the time.

Anna's junior high school offered a geography club, led by an enthusiastic teacher who convinced Anna's mother that she needed to go away to college "because she knew how sheltered the home was." And so Anna went to a state college where her teacher had some ties and majored in geography. "About eighth grade, I [had been] to a conference and heard researchers talking about research in Ecuador. I was just completely fascinated, so I always kind of felt like I wanted to go on, and I wanted a Ph.D."

Like Pam, Lynn Becker had an interrupted education following her marriage at age twenty. Her early years had given her a strong dedication to foreign-area studies.

> My father was an old radical. . . . He fought in the Spanish Civil War, then became involved in [other international political activity]. He wanted to take us to Africa. As I child I started reading books on Africa, and I believed we were going to go. Unfortunately, my father was an alcoholic and had certain other problems, and we didn't go anywhere. . . . But I just couldn't get unglued from [my interest]. . . . One falls in love with a topic . . . and it just sort of holds you.

Lynn's father had dropped out of high school, but her mother, a teacher, had graduated from college. Both parents supported the idea that Lynn should go to college. She enrolled for a year and a half in junior college, but dropped out to marry a graduate student studying on the other side of the country. They had a baby right away.

Elaine Phillips and Chris Jenkins had perhaps more "conventional" middle-class childhoods for their cohort than Ellen, Pam, Anna, or Lynn, with the expectations of their class that they would attend college and possibly become schoolteachers. Elaine's father had stumbled into a career as an accountant in a large company in a medium-sized town. Her mother was a housewife who later worked part-time when the children left home. "In those days, coming out of the kind of town I did, and having the kinds of friends I did, [teaching] was a job you had to go into. I wanted to go to college, but you were looking for something that you could slip in and out of and have a family at the same time. Just all those horrible old things that we used to say."

Elaine enrolled in the major university in her state in an interdisciplinary liberal arts program. She quickly decided education courses were not for her, though "I think teaching was still in the back of my mind. But it took a long time to ever get the idea that college teaching might be the place for me." Taking a geography course to fulfill a science requirement, she encountered a woman teaching assistant who encouraged her and indicated that geogra-

phy offered career possibilities. Elaine was "dating a young man . . .
who decided to go into the Peace Corps, and for a while I was going
to go with him, until I realized I was going to end up in Africa, in
the Cameroons, and that's really not me, I guess. And so I decided to
go to graduate school."

Chris Jenkins's parents, teachers in a small community, valued education a great deal and thought the local high school quite inadequate. They enrolled Chris in a girls' boarding school that "was
extremely demanding academically and all of our teachers were
women . . . who nowadays would be pursuing academic careers at
the university level. . . . I think that place really changed my life."
She went on to a liberal arts college, took geography in her sopho-
more year, and connected it to ideas that had enthused her in high-
school history classes. Her career aspirations were vague. "I just
wanted to survive to the end [of the B.A.] and wanted desperately to
go into the Peace Corps. . . . [I]t was absolutely the only thing I
wanted to do. . . . A lot of our friends went into the Peace Corps
. . . it was viewed as a good thing to do, an exciting thing . . . a creative thing." Chris married as an undergraduate, and went to Africa
with her husband as a teacher. There her first child was born.

Redefining the Self: Marriage, Motherhood, Education

Though differences in the values and experiences of their families, especially issues of class and links to other women, emerge in
different ways in the stories of these women's early lives, at the point
of marriage and motherhood their stories begin to diverge further,
with implications for their professional futures. They also begin to
talk about themselves in new ways. Differences in relationships with
husbands and the women's interpretations of their roles as mothers
are key, to be sure, but these also intersect with changing conditions
in funding support for higher education and the job market. It is
therefore critical to examine the issue of timing and context in the
women's professional lives.

Chris's husband encouraged her to apply for a Ph.D. program
rather than an M.A. program after the Peace Corps: "You may as
well go for a Ph.D. . . . you're more likely to get a fellowship if you
say you want to go for a Ph.D." Both were accepted into the same

program and she gained a National Defense Foreign Language fellowship. That "actually turned out to be a godsend, because we had to take four courses . . . we had a kid . . . and if I had been taking hard-core economics instead of the easy language course, I would just have sunk." As a married woman with a young child, she was viewed as strange: "Well, I was strange. I am weird. I've always been weird. . . . And most of the faculty . . . really didn't know what to do. . . . My advisor really didn't have a clue how to deal with me. . . . His basic view was that I should be home with my kids [She had a second child after her second year]. . . . Whenever I would talk with him, the conversation was always about children and family, and never about anything professional."

Chris felt she really had to prove herself in graduate school. In the first year, she was the only student in one or two courses not to get an incomplete—then "they knew I was a serious student." Chris obtained a National Science Foundation grant to undertake dissertation research abroad and she and her husband worked on complementary projects. Graduate school thus became a point where Chris, always an eager student, responded in ways that show some transition in her expectations regarding professional work, no longer restricted to "substitute teaching." Nevertheless, she presents her self as naïve: "I figured I'd get a job somewhere. I wasn't too worried about it. I was unbelievably naïve."

Lynn's and Pam's approaches to motherhood and education followed a different path, with long-term consequences. When they became mothers, both withdrew to the home. As their children grew, each felt the need to seek more education (likely influenced by the 1970s culture of the women's liberation movement) and enrolled first in a community college, then in large metropolitan universities, completing her graduate degree in the same department where each took her B.A. Their choices of institution were shaped by the locations of their husbands' jobs, not because the schools necessarily best fit their needs. Pam had two children when she returned to college.

> It was really tough, because I felt like I was the oldest person . . . but I got used to that. . . . The toughest part was the home situation because my husband had never expected that I would go to school and

become very gung-ho about what I was doing. He had expected that, "oh, she's going to take a few classes. She needs something else now." . . . I became so enthusiastic that I probably overdosed everyone around me. I had completely changed. I had become more assertive. . . . And I began to have my own opinions that were a lot more liberal than my husband's or my family's. He didn't adjust for a long time . . . five or six years were really terrible. I felt really trapped that the marriage wasn't working but yet I didn't have a way to support myself. . . . I knew he wasn't happy, but yet I never said, "as soon as I get my degree I'm going to get a divorce."

Despite her claims that managing study was "simple," Pam rose around 4 a.m. to study. "My kids learned to do their own work, as far as washing and things like that went. . . . My daughter thought it was really neat. My son never quite thought this was a neat idea . . . he always wanted a mother to stay home and bake cookies." Dealing with family crises was "nerve wracking. . . . When something bad happens [and they are young] you don't just say, 'I have a test tomorrow.'"

Pam also made adjustments. She had an opportunity for a short period of international fieldwork and became interested in pursuing work in that region. "But . . . I felt my options . . . were limited because I had a family and I just couldn't take off. If I could have done what I wanted, I would have been a person who did much more fieldwork than I have done." Instead, Pam chose a dissertation topic that could be researched through archival work close to home. Her choice also reflected her thinking about her age. She was eager to finish in her early forties, and get into a career. The local dissertation could be completed more rapidly than a foreign project.

For Lynn, entranced by a foreign area since childhood, changing topics was not a choice. She persisted in finding ways to support overseas travel. It has been extremely rewarding to her personally, but also very difficult. She had serious guilt feelings when she first left her family for a weekend's local fieldwork. "I think I probably would have done more physical geography . . . but when the children were young it was very hard to get away for weekend trips." When she first went away "I had terrific nightmares. A lot of separation anxiety. But it worked out fine . . . my husband's a perfectly capable individual. They enjoyed the weekend. But I was the one with the problem."

Despite her anxieties, Lynn planned her study abroad, leaving a notebook, "How to Survive while Mother's in Ghana." "It was one of the few times that I said, this is something I want, even though it's not necessarily good for the family. It's something I want so badly that I'm willing to override my guilt and my feeling that everybody else should have what they need first." Her husband was really supportive about it, so she went. Study at home was also stressful.

> My life was just crazy, frantic, and I would drop from exhaustion once in a while. My husband's mother was very angry that I was going to school and not staying home and taking care of her son and children. I got a lot of pressure from her . . . but she also showed up at a couple of critical moments and took care of the children. . . . I've come to realize that most of the conflict was going on in me.

Unlike the women who entered graduate school at younger ages and when fellowships were more available, both Pam and Lynn worked (in the university and outside) to support their educations, neither feeling it was her right to draw on family resources. When they held teaching assistantships they experienced stress from fellow graduate students who thought that married women should not be receiving support. "I got called the housewife geographer, which made me so mad," said Pam.

Neither Ellen nor Anna had children until they had faculty positions. Ellen's delay was connected with her desire to travel and do fieldwork with her husband, who was a graduate student in a related discipline in the same university. When an opportunity came to do field research on a professor's project in a remote foreign area, they grabbed it.

> People with better sense would never have considered it. . . . And we just loved it. We just knew this is what we want to be doing for the rest of our lives. . . . And I also felt very optimistic that this was going to work. And it felt that way for several more years . . . [W]e had planned to have children, but just didn't get around to it, because it was never the right time. We couldn't afford it and we didn't want

to have a child in the jungle. . . . [But still, w]e thought, we'll have a kid, we'll get teaching jobs, we'll have these dissertations done, and we'll be on our way.

Anna was almost finished with her degree and beginning to look for positions at the time she married a fellow student. Rather than be separated from her husband, she deferred completion of the dissertation for a year in order to join him abroad where he was doing dissertation field research, opting out of the job search. Both returned and finished their dissertations the following year. She did not have children during this time.

Elaine entered graduate school with support from a National Defense Education Act Fellowship. She hadn't really thought much about whether to do a Ph.D. or a master's, but the fellowship started her on the doctoral track. "I guess this whole thing was given to me. It was kind of dumb not to go on for a Ph.D. I guess I always had the feeling I could back out at some time along the line if I didn't like it." In her second year of graduate school, her future husband entered the program, also on an NDEA fellowship. Within a year, they married. "So that changes your attitude towards a lot of things— changes your thoughts about supporting yourself and all kinds of things like that." I asked Elaine if she thought departmental attitudes changed toward her. She found that difficult to evaluate: "It's all a long time ago. I don't think they took me as seriously as a student and as a professional after that. I was always a good student . . . but it's the drive, I don't think was there in their minds. And I don't think it was there in my mind at all in this whole time period." Elaine was able to complete the research for her dissertation before her daughter was born, "so I did continue through with this whole process, but not at the same pace and not with the same fervor that I started the whole thing." Writing the dissertation, however, stagnated for several years as her husband secured an appointment. She then became "a young faculty wife" and mother of two children.

The Job Search and Beyond

Chris, Ellen, and Anna all completed their dissertations in the 1970s, at a time when affirmative action was just coming onto the academic scene, but also when the booming job market of the late sixties had passed. All three were married to fellow academics. In the early 1970s, Chris and Anna and their husbands began sending out applications separately. Though neither Chris nor her husband had quite finished the dissertations, "we both sent out résumés to a lot of places and [he] always got the response, right? So he got a job . . . and the second year we were there, I got a job half-time in another department and half-time in geography." Chris has no recollection of how the position was created: "I was so naïve then. I had no idea."

Chris and her husband have subsequently made several career moves, initially because of tenure problems, but now both have established, stable, senior positions, though in different institutions. This has been possible because he has been able and willing to change fields whereas she remained in geography. For a time she supported herself with grants and teaching the odd course at the institution where her husband was employed, but knew she needed to get out. By this time she had also begun to recognize structural problems for women in academia. Today, she describes herself as more socialized, "though I'm sure there's still a long way to go."

Chris sees her productivity as initially low, partly because of having young children, but also because of the mindset instilled in her by her graduate program, which "was not conducive to one's self-confidence and self-esteem, and the idea that you really had something to contribute." She also identified a lack of mentoring and has worked consciously to be a mentor to her own students. Over time, her productivity increased, but the changes she sees are not only in work, but also in gains of personal time—for reading and exercise.

Ellen's husband has also accommodated to her career, though at costs to his own that have magnified over time. "We applied for various jobs, and we were so naïve. We looked at the job ads, we didn't have any sense of networking, and still felt sort of intimidated by the whole thing." And we didn't think "that we were not going to get two jobs in the same place, and that I should not be getting pregnant. But nobody ever talked to us about that." Ellen's first job offer came from a graduate department, and she took it. At that time and place, spousal placement was not part of institutional thinking, so her husband was ignored by the institution. He stayed home, took considerable responsibility for household and childcare, and worked on his dissertation. When she finished a little ahead of him, tensions increased. Though "he was very happy for my career to come first, and was not concerned about what status institution he might go to, he did want to teach and do his research." After attempts to locate other positions—"anywhere we could be together"—and a period of leave when Ellen joined him on a campus where he had an opportunity for a continuing job, they have now experienced a decade with her husband in on-and-off, part-time, temporary appointments.

Anna first became aware of discrimination as the job searches progressed. "I went to a school . . . that was extremely interested, until one of my referees wrote that I had taken nine months off and gone to [field area] with my husband. And then the tenor of their letters changed." Though gaining field experience (albeit related to her husband's project), she became marked as "a wife," rather than as a serious professional. She continued and "had another interview [at a state university] and he [the chair] was real interested. A year later I met him at the AAG convention with the person he had hired. And he said to this man, 'I really want to hired her, but I couldn't live with the fact that she and her husband would be living in different states.' "Anna resented his paternalism, and also that he would make such a remark in the presence of the man he had hired. Anna's husband obtained a short-term position and she tried unsuccessfully for a job nearby, then joined him. "His colleagues treated me as if I didn't know anything . . . they just treated me as an extension. One of the wives said, 'Well, if you can't get a job here, do what I did, grow a baby.' And I thought that was ridiculous." Her experiences illustrate prevalent attitudes of the period—that married women did not belong in the profession, were not perceived as seriously committed, and should not expect to be professional geographers, and that motherhood should be the alternative.

Anna continued to interview but was also prepared to consider other options, such as entering law school. Unexpectedly, she was offered a tenure-track position within commuting distance of her husband's job. She was very happy with this department, with her teaching and research, and especially with the way the department accommodated her when she became pregnant and had a child. Her husband was strongly oriented to pursuing a research career, however. When he was offered a tenure-track position in a distant state, he made a largely independent decision to go and pressured Anna to follow. She places high value on family togetherness, remembering her own childhood in an ethnic community without a father. So she resigned and joined her husband. Since then, her life, both personally and professionally, has been difficult. She displayed flexibility finding positions outside academia, but felt her husband did not respect her work. He also left all the housework and childcare to her as he pursued tenure. Then he became involved with another woman and he and Anna divorced. Because she wanted to keep her children near their father and because her mother had also moved to the same community, she has remained in the same city, piecing together consultant work and part-time teaching in the restricted job market. She knows these positions do not build a curriculum vitae that would allow her to reenter university in a good academic position. She says, "I wanted us to stay together, but it didn't work, even making the sacrifices in my career. If I had any of it to do over, I'd probably do it all the same way except for the sacrifice of my career."

For Pam, the stresses she felt as a mature-aged student and mother modified as her children grew and she and her husband developed an understanding of their goals. While finishing her dissertation, she obtained a temporary position in a state college within commuting distance of their home. When it was converted to a tenure-track line but opened to a national search, she applied but also "applied all over the country" with her husband's support. "Just do it, and then if you get one, we'll decide. I haven't watched you do all this work not to get a job at the end," he said. She secured the local job and has begun contemplating international field research: "I'm at a good stage in my life. I've done the children thing, and now I have a career to do what I want." She sees herself as somewhat out of sync with her husband, who has begun to think of retirement, whereas she is at the beginning. Her work has, however, "opened up a new world to us," bringing new friendships and interests such as travel that he now shares.

Lynn's experiences in searching for a job have been frustrating for three reasons: wanting to stay in the same area as her husband (now a senior academic), wanting to practice her own regional specialty (for which very few positions have been available), and, especially, being her age. She encountered several rebuffs in seeking fellowships and jobs because of her age. At one AAG meeting, on being introduced as a potentially strong candidate to a senior geographer in her field, the first thing he said was "you aren't twenty any more, where have you been?" She feels that "older men think you are as tired as they are." Lynn has decided to widen the area of her job search and also to seek positions in related disciplines but meanwhile is working in an administrative job, taking whatever opportunities come to get back to the field, and feeling increasingly depressed as time goes by.

Elaine, whose appointments have also mostly been part-time, has made a different path with which, after some struggle, she has come to terms. She has taught part-time in her husband's department, done editing work and some writing that have yielded royalties. Elaine has also contributed to her husband's career. "There would always be times in our lives when I would say 'I've got some time, is there anything that you want to know about for your course?' " She has come up with ideas for papers for which she has done a significant part of the research and has even written the first draft. She did not claim senior authorship because "he had some grant money to pay for part of this" and needed to report his productivity to the funding source. Elaine's consciousness shifted dramatically when she was offered a one-year appointment at another institution during her husband's sabbatical and encountered lively feminist colleagues. The year's experience was exhilarating, prompting her to take up a major research project. She hated to come back to the former situation. "There was a time in my life when I was really angry about this." Now, however, she says: "I think as I get older, I've gotten much more mellow about the whole thing. Forget it. It's not worth that amount of energy." Shortly after our interview, Elaine wrote:

I've thought about our interview several times . . . wondering if there's anything I want to change. Instead, I want to emphasize that I'm quite happy now with a minor teaching role and research/publishing in areas where I want to work, albeit topics that will result in royalties. I'm getting a little too old to go the assistant professor route—I'm not "hungry" enough. I suppose if I had to pinpoint one area I'd like to change in this academic business is that decisions are made on a male's life calendar. We're losing too many women in the tenure years that are unfortunately the childbearing years. Around 40 a woman is ready to return full-time, mentally and physically. Part-time opportunities in the intervening years are the answer.

Branches in the Road

In reflecting on the stories of these six women geographers who are at midlife, I have been struck by the fact that they represent a transitional generation even as they made personal transitions. They came to adulthood at a time that the "women's liberation movement" was beginning to challenge gender ideologies and when the economy was fostering increasing participation of women in the labor force. They did not face the same constraints as academic women geographers of earlier generations, who were mostly channeled into careers in teacher-training institutions or women's colleges (Monk 1998). Nor did they forgo marriage and motherhood as earlier professional women had done. Still, they were not free of post–World War II values that supported women's college education while assuming that their career expectations would be limited and that marriage and motherhood would be prioritized. Their world has not been one in which talented young women were encouraged to plan careers, seek mentors, or develop expectations (though not necessarily enjoy the realities) of 'family friendly' policies in institutions. Only Anna expressed high professional aspirations at an early age, though Pam, Chris, and Ellen represent themselves as "different," "weird," or "independent." Throughout their narratives, the two who have attained the most elite professional status, Ellen and Chris, say they are "naïve." And as Ellen says "I've never had a sense of what I would be doing ten or thirty years down the road. But I've made decisions at each point that with hindsight lead into each other."

Within their context, transitions in thinking about the self and career have come at different and multiple points in the life course, reflecting different personal relationships. Ellen credits two great aunts with fostering an independence that her parents were not promoting. Chris and Anna speak of women school teachers changing their lives, but only later did Chris come to a sense of awareness of "proving herself serious," of recognizing structural problems for women in academia. Today, Chris still describes herself as "not fully cooked" in relation to mastering academic policies and strategies. For Pam and Lynn a sense of "doing it for themselves" came in their late twenties and thirties, as they returned to school following early marriage and motherhood. Elaine experienced a period of heightened ambition after encountering feminist academics, but later "mellowed" and found satisfaction in prioritizing marriage and motherhood over full-time academic work.

Though my questions prompted discussions of the relationship between marriage, family, and professional life, the narratives indicate that it is unlikely that these women would have been able to describe their careers without sustained reference to their adult personal lives. This is in contrast to those men whose autobiographical accounts are in the literature. These women acknowledge what it has meant to juggle the two-career household and how their husbands' careers have been intertwined with their own, whether for better or worse. Indeed, the husbands' adaptability, flexibility in career expectations and options, and support (or lack thereof) are repeatedly acknowledged as key influences on the women's professional opportunities and experiences.

I have not dwelt on the substance of these women's geographic research and teaching because it is difficult to do this and maintain the confidentiality I promised. In the context of an interpretation that invokes cohort analysis, however, it is noteworthy that all of these women came to geography in a decade when feminist consciousness was entering geographic scholarship. All to some degree expressed this consciousness and have included gender content in their teaching or research or both.[6]

My reflections on these women's stories lie less with issues of the creation of geographic knowledge and more with the social construction of the profession and its institutional settings. Ellen, Anna, Chris, and Elaine completed graduate school when the themes of gender discrimination and the status of women in the profession

6. I knew the feminist work of some of these women prior to the interviews, but in other cases learned of it during the interview.

were just emerging in the literature (Berman 1977; Rubin 1979; Zelinsky 1973a, 1973b). The assumptions, language, and practices of men they encountered in their job searches were not generally sensitive to the recently legislated affirmative action policies. Though the women did not discuss antinepotism policies, these may well also have affected their situations. For Pam and Lynn, going on the job market several years later, the potential of age discrimination loomed larger than that of gender barriers. They did not fit the model, built on assumptions about the male life course, of the bright, young professional, single-mindedly pursuing a career.

Though public rhetoric and some policies have changed in ways that enhance opportunities to combine personal and professional lives, for example, with respect to spousal placement and reduced duties during pregnancy, these changes are fragile. Further, old assumptions persist about rapid productivity in the early professional years for those hired into tenure-track positions. Increasing reliance on a contingent labor force of part-time faculty also offers no long-term panacea for women juggling caring responsibilities with a career. Such work does not allow one to build competitive credentials for privileged positions. Nor has thinking about combining the personal and professional generally moved to consider the roles of later midlife. It remains a challenge to create alternative pathways.

 Engaging Autobiography

Pamela Moss

A Tentative Review

The contributors to this collection highlight various explicit uses of autobiography in geography. Although only a few talk about what it is *like to write* autobiographically, all *write* autobiographically. Some were confessional, others more harbored; some focused on life story, others on process; some revealed the details of their own lifeworlds, others used reflexivity to position their research lives. Some wrote about their own lives, some wrote about their lives in conjunction with others, and some wrote primarily about the lives of others.

Rather than go into a detailed reading of each of the chapters and tell you what I wanted you to get out of each, I leave the interpretation up to you. For in the end it is really up to you, the reader, to figure out how each author engages autobiography and how successfully each writes his or her life. Instead, I prefer to give short introductory renderings to initiate the interpretive process. It seems only fair that I should give my take on each of the contributions, sort of a "first cut" on interpretation, in order to abet the discussion of critical uses of autobiography in writing geographers' lives. After all, I was the one who wanted to edit a collection on the topic. For my framework, I rely on the three ways geographers use autobiography that I noted in the first chapter: to chronicle the history of the development of the discipline; to approach research; and to analyze information gathered through the research process. All the contributions here can be read through these categories.

Chronicling the history of geography entails addressing not only the literatures emerging from geographers' pens and printers, but also the geographers themselves. Buttimer, Eyles, Archer, and Monk address the issue of chronicling the discipline through the documentation of careers. For Buttimer and Eyles, long-standing, senior scholars who assisted geography in opening up to humanist perspectives, opposing disciplinary prognoses emerge when I juxtapose their thoughts on their careers in the discipline: Buttimer finds geography dynamic and flourishing whereas Eyles thinks its limits have been met and its demise is on the horizon. For Buttimer, place is central to the unfolding of her career. The shifts and changes in her life are linked to the places she's lived and the connections she's made there. Each and every research project is linked to the world surrounding her—Seattle, Worcester, Louvain, Glasgow, Lund, Ottawa, Dublin. But she isn't place-bound. Her efforts to create an international community in geography have been successful, and she is being rewarded for these successes.

Eyles recounts his career in terms of intellectual moments within geography. In disclosing the internal links of his career, he moves through his interests while tying them to the general trends in geography over the past thirty years. He seems always to have been pulled into interdisciplinary work, first in social policy and then in health and environment. His dismal foreboding for the course of geography is wrapped up in his ongoing unease with big T Theory. Geographers are now moving through a moment where big T Theory has brought on the creation and dispersal of small t theory—small t in terms of specificity as well as of the post-structuralist notion of resisting grand narratives. Ironically, as we can see from his documentation of his career, Eyles is partially responsible for the movement that he now fundamentally rejects.

Archer's chronicle is also linked to place through other people, in particular, his life partner, Ingrid Bartsch. Part and parcel of the development of his career are the many moves they've made over the years. Archer contrasts his and Bartsch's career path with the omniscient, idealized standard of the "academic's" career. Recounting in detail the making of a (dual) career generates insight into the innumerable ways different aspects of our lives constitute one another. He makes the point that our autobiographies are not only ours; they are also accounts of those people close to us and of the world surrounding us. Interestingly, too, he leaves open his own interpretation, leaving Ingrid to write her own story, her own take on the events, her own autobiography.

Monk draws on the autobiographies of women pursuing careers in geography (collected as life stories). She grounds these women's careers in the historical milieus within which they emerged, including the situations that both constraind and enabled the women's choices—academic ability, marriage, job interviews, family life, household moves. Her own life history, recounted at the beginning of the chapter, indicates that her pursuit of a career, too, was grounded in her lifeworld in Australia. In juxtaposing these career accounts, she implicitly paves the way for alternative future career paths and shows how who we are can have an impact on the choices we make about our careers.

Although qualitative methods have been increasingly popular in geography, autobiographical approaches are not widely accepted as a standard method for geographic research. Yet engaging in reflexivity in order to position oneself in the research process has gained acceptance in geography.[1] Roth and Saltmarsh push the boundaries of reflexivity as part of a range of accepted methodologies. Roth initiates a reflective discussion about her motives in wanting to do development work. She problematizes not only the notion of development, but also much of the language associated with it, all within the context of her plans to engage the world around her. The tensions she identifies and tries to work through are useful in understanding the process through which she came to decide to pursue development work. Although it appears that the introduction and the "conclusion" are the same, in that she decides that she can only be herself, the path she forges through the literature assures us that this is not the case. Although she seems to be a little more confident at the conclusion than at the introduction, she also seems to

---

1. Rose (1997) sees this differently. She argues that the literature has created a discursive feminist geographer who can see everything and fully understand her own positionings. (This argument has yet to be challenged in print.) She cautions against this type of reflexivity and urges the recognition of "absences and fallibilities while recognizing that the significance of this does not rest entirely in our own hands" (Rose 1997, 319).

recognize that she has no closure on her choice of approach to research, and perhaps even that there isn't any.[2]

Saltmarsh faces head-on some of the most basic and deriding arguments against writing autobiographically in the academy. As she lays out her approach, she weaves her reasons for asking the questions she does together with where she thinks she needs to be to answer them. She makes an impassioned avowal of being "who I am" and doing research "as I am" in a place that ranges from the inhospitable to the antagonistic, from cool to hostile. Even so, she is tenaciously determined to proceed, valuing where she has been and acknowledging its impact on her future.

On the basis of autobiography's tenuous standing as a methodological approach, it seems unlikely that autobiography would be accepted as an *analytical* method. Yet Butz, Cook, and Knopp variously use autobiography to build an analysis of their research interests. Butz, reluctant to give up Habermas's theory of communicative action in the face of poststructural readings, shows how he is wrapped up in the constitutive process of constructing the lifeworlds of the people with whom he carries out his research as much as they are in constructing his. Through writing autobiographically, he explains in detail how discursive practices recursively shape him, his family, and some Shimshalis. His account of this interactive process, or "bleeding," as he calls it, shows how entwined his life and the lives of the people with whom he conducts research are as they engage the world around them.

Cook, too, writes autobiographically to construct an explanation of why and how he almost did not receive a doctoral degree. In calling into question the established ways to complete doctoral studies, he challenges the very notion of being an academic in the academy. Through readjusting, revising, reflecting, rewriting, and rethinking his autobiography, Cook himself emerges as an analytical tool to reanalyze the information he gathered, the way he gathered it, the way he presented it, and the way he wrote it up. He uses his graduate school experiences analytically, critically assessing the content of the research and the research process itself.

2. See also Davis 1997 for a discussion of how theoretical tensions can be effective in understanding how theory and practice intersect.

Knopp situates himself personally and professionally by sorting through important issues he has faced since childhood. Now poised at the intersection of an institution and a community, he brings together his passions and shows how he spans both. He uses his insights from the literature and puts them into practice; he draws on his experiences and theorizes them. By "activating" queer politics, he is able to establish practical links between queer abstractions and queer social practices. He uses his life, his autobiography, as an ongoing analytical framework through which he makes sense of his everyday life in the academy and in the community.

This categorization of using autobiography to chronicle the discipline, as a methodological approach, and as an analytical method seems to be useful. It places these contributions in the literature in some way, giving them context (albeit my context), and it begins the interpretive process. But my take on these contributions is just that, my take. Sometimes I don't think it's necessary to say that what I write is not definitive, is not the only interpretation, is not the "Truth." But sometimes I think that it is. If I don't, then you as the reader might think that what I write is either *the* only way to interpret these writings, or that I think this is the *only* way to interpret them. It's not. Had I written this chapter three months ago, three weeks ago, or three days from now, my interpretation would be slightly different. I can imagine that it would generally be the same, at least at this point in my life, but I would probably use different expressions, words, and phrases. As you read, too, your interpretation will vary depending on the context within which you read these contributions—for a course, for interest, for research, for criticism, while you're hungry, while you're sleepy. And, as time passes, we may both abandon our initial interpretations and favor those more in vogue, more critical, more in tune with our current lifeworlds.

But this is not the only thing I want to say by way of concluding this collection. Just as I'm reluctant to offer my reading of the contributions here as definitive, I'm hesitant to leave off without any synthetic comments on the collection as a whole. In lieu of another reading of the contributions, I choose to tie them together through three tensions I find within the collection itself. The first deals with the imposition of disciplinary boundaries; the second, with not knowing the contributors even after they have written about themselves; and the third, with the incompleteness of autobiography.

Some Remarks about Geography, Knowing, and Incomplete Spaces

Having read and reread the collection several times over, I'm still ill at ease with the disciplinary boundaries I (purposely) set for the project. Over and over again, geography is discussed as if it were a bounded social science with clear directives for undertaking geographic research and for being geographers. In a sense all the contributors (myself included) are bounded by geography in some way. I set the literature review primarily within geography just as I chose to invite only geographers to be contributors. Buttimer and Eyles firmly ground themselves as geographers in their training and in their teaching, even if they diverge in their assessments of the usefulness of geography *as a discipline.* Archer and Cook talk about the restrictive borders of *departments of geography:* how administrative units within the university are powerful in delineating who belongs in which discipline and in how departmental members assess the quality, scope, and content of research. Knopp, Monk, and Saltmarsh discuss notions of what *being a geographer* is all about—choosing a "queer" path in geography, linking women geographers' lives, and doing geography as one *is.* Only Butz and Roth appear to be free of the confines of geography's disciplinary boundaries; neither place geography centrally in their discussion. Perhaps they are more closely linked to interdisciplinary aspects of development work. Perhaps they read widely outside the development geography literature. Perhaps they don't think of themselves as geographers. But this absence in writing doesn't reflect their material grounding in the discipline: Butz trained in geography and is a member of a geography department and Roth is pursuing a doctorate in geography.

How do we break these binds? Or do we even want to? In some senses I agree with what Eyles forecasts for geography, at least human geography, as a discipline. Sometimes we get caught up in asking "why is this geography?" and "where is the space?" when perhaps it doesn't matter, at least in some contexts. I'm wondering if autobiography is one of these contexts. Some time ago I read James Ellroy's autobiography, *My Dark Places* (1996). In it he recounts the story of the investigation of his mother's murder—an unresolved sex murder from 1958—and provides us with glimpses into his life as a white male growing up in 1960s Los Angeles. Yet most compelling for me was not his extensive descriptions of the urban environment, or the intensity with which he recounted the murder investigation and his petty criminal forays as a teen, or the detailed characterizations of the people who were part of his life—all of which were remarkable. Rather, what struck a chord with me was his comment that while America's eyes were riveted on O. J. Simpson's courtroom, just down the hall the verdict of *another* trial of *another* sex murder was *yet again* being handed down. This simple comment pulled together sex, class, violence, and status for me in a way that made more sense to me than ever before. Like Ellroy's mother, the young woman, neither famous, nor rich, was sexually assaulted, then brutally murdered. And while Americans watched a spectacle, the very same things that happened to Ellroy were happening in someone else's everyday life, just as they had been happening in the lives of others for years on end. Ellroy's autobiography was like Cook's variety of autobiography—"it-me-them-you-here-me-that-you-there-her-us-then-so"—connecting, de-linking, and reconnecting moments and people over and over again through specific spaces—Los Angeles, courtrooms, news media, police headquarters, bungalows, highway exchanges, vacant lots, hometowns.

Was Ellroy's autobiography place-sensitive without making universal claims? Definitely. Did Ellroy provide insights into the way American culture deals with violence collectively? Of course. Did Ellroy sit around thinking about how he was going to spatialize his story in these particular ways? I don't know. Perhaps autobiography inevitably brings with it notions of space and place. Perhaps space is already always present in autobiography. If this is the case, then maybe this is a point where I can agree with Eyles about the demise of human geography, in the sense that a spatial understanding is fundamental to any social relation. Perhaps we don't need a discipline to make this point over and over again. Perhaps autobiography is just another avenue to make this point.

Having read and reread the collection several times over, I'm still unsure whether I know these contributors. I feel that I want to know more; I want to ask questions; I want to engage their autobiographies. I've never met Buttimer, Cook, or Saltmarsh face-to-face. Even though our e-mail exchanges are full of our calling each other by our first names, I'm not sure that I will be able to call them by first name when I meet them. Other contributors—Archer, Knopp, and Monk—I know through various conferences I've attended. The rest, I know a little bit better. Butz and I studied with Eyles at McMaster. Roth took classes from me in Victoria. Even with these connections, I'm left with the feeling that I want to know more. I want to know how the careers of Buttimer and Eyles are linked with their "personal" lives. Though Buttimer makes mention of her partner, this aspect of her life is not developed in this piece. Eyles doesn't make mention of his personal relationships, except when they are part of his professional activities. I want to know how Saltmarsh and Roth are doing in their doctoral programs and how they are living their lives as students, and if they are resolving the issues they've problematized in their chapters. Just from the way Cook used his humor and the pace of his writing, I look forward to encountering the animated version of him—"Cook Unplugged." I'd love to see Archer in person using his integrated notion of autobiography and Butz using his notion of "bleeding" autoethnography to show how other people fit into and are a part of their lives. And I also want to have coffee with Monk and talk about women's academic careers and perhaps teach a course with Knopp on community activism as research, if I couldn't take it myself.

Would these acts help me know these people better? I don't know. The more I interact, however, the more I think that our autobiographies would have to become part of one another's. Thinking through these connections and how lives become more entwined over time, I'm reminded of a piece that Nicole Brossard wrote, "Green Light of Labyrinth Park, La nuit verte du parc Labyrinthe" (Brossard 1995). One night the narrator goes into a labyrinth with many other women. At each turn, she comes to know a little more about herself and the other women. At each bend, she links her feelings with what she knows about language, politics, lesbians, women, and life. She is drenched with emotion and creativity. When she emerges, the spell is not completely broken. While she thinks about how these women have to return to their homes and engage their familiar surroundings once more, she knows that each will have with her the memory of the walk in the labyrinth one solstice night. I'd like to think that writing one's life as a woman, while engaging autobiography, is like Brossard's labyrinth. I'm not sure if such an analogy will help me in knowing these people any better. But I think that it will help me in knowing how to know them a little better.

Having read and reread the collection several times over, I'm still uncertain about how these chapters fit together. I don't think that they are a series of non sequiturs. There are themes that link them together. I don't think they represent all types of autobiography or its uses. They are limited in scope. I don't think that they are either too theoretical or atheoretical. They engage autobiography theoretically, yet don't theorize autobiography; they draw on life experiences, yet don't remain mired in empirical data. I think, for me at least, it is more the incompleteness of the chapters, both individually and collectively, that leaves me uncertain of their connection to one another. There seems to be no doubt that any autobiographical writing has to be incomplete: we simply can't know everything about our lives. And we also know that we are not isolated events— we are connected to other people as well as to society more generally. Yet I'm overwhelmed by my wondering how our lives are connected to one another's and to our surroundings. What links are not a part of these chapters? How is it that "who we are" juxtaposed with "who she is" matters? How will writing one's life assist us in coming to terms with our connections to one another?

Do these chapters have to fit together? At least to some extent, yes. Do I have to elaborate how they fit together? Probably not. One of my favorite books is a collection of essays on home (Abraham et al. 1991). As I read each essay, I pictured each woman's home, even the parts they didn't describe. I created elaborate pictures of their living spaces through which I deluded myself into thinking that I knew them, that I was close to them somehow, that they were part of my life. I quickly chastened myself for assuming such intimacy. I couldn't know them; I could only know the parts that they chose to reveal to me through their writing. And even though they have no clue who I am, let alone that I exist, except in a disembodied sort of way, these essays are very much a part of who I am. They shaped how I came to see my home. They are part of my life.

When engaging feminist and socialist praxis, I have these issues constantly on my mind. I wonder how my training in geography entwined with my other life experiences can contribute to, for example, constructing community housing projects for women leaving abusive relationships or assessing living space needs for street women. Working collectively for me means not only being me, but also trying to figure out who I'm working with—the people whom I know only a little about, but about whom I want to know more. This incompleteness of the visible and conscious paths of the connections to other people makes me wonder how we should approach one another. Do I organize on the basis of common political goals or seek diversity through coalition politics? Do I focus on similarities, differences, or some combination? And how can I do this autobiographically? Must I only act on connections that are visible to me and to other people? Or, can I act and sort out the connections later? Do I engage the world around me as I am? Do I have to know who I am? Do I have to know those with whom I'm connected? In short, how do I traverse incomplete spaces?

If I had to write the "Autobiography of Pamela Moss" again, I don't know exactly what I'd include. I guess I would begin where I left so many years ago:

I didn't see my grandmother often while growing up. I was never sure that she even liked me. There were so many grandchildren, and I was in the middle of the pack. It was just like my own family, being in the middle of two brothers, only with more siblings. But when I broke my arm she sent me a dress—a red maxi with white rick-rack. I wore it until the edges raveled and the thread disintegrated.

It wasn't until four years later that I flew in an airplane—from Indianapolis to Washington D.C. It was for a school field trip. I eventually had pets when I was in high school, a blue parakeet with squinty eyes named Clint (who flew away) and his replacement Cornelius (who died of a heat attack). I also had two rabbits, Cicero, a beautiful, black, house-trained bunny, and Nutmeg, who lived the last few years of her life with only one hind leg. I didn't move away from the house on Brookside Lane until I was twenty-one, to Vancouver, British Columbia, to go to graduate school. This was just after my dad died of lung cancer. He smoked. My grandmother died in 1987 after a long illness, probably Alzheimer's disease, though it's not completely certain. I have a nephew and niece, Samuel and Hannah. I'm enjoying watching them grow up.

How would I write the rest? I don't know. There is just so much. All I know is that it would be full of other people—my mom, my brothers, my friends, my colleagues, my coworkers, strangers. It would be about who I am and how I engage the world around me. It would be about coming to know who other people are. And it would inevitably be incomplete.

Works Cited

Index

Works Cited

Abir-Am, Pnina G., and Dorinda Outram, eds. 1987. *Uneasy Careers and Intimate Lives: Women in Science 1789–1979.* New Brunswick, N.J.: Rutgers Univ. Press.

Abraham, Laurie, Mary Beth Danielson, Nancy Eberle, Laura Green, Janice Rosenberg, and Carroll Stoner. 1991. *Reinventing Home: Six Working Women Look at Their Home Lives.* New York: Plume.

Abramson, Allen. 1993. "Between Autobiography and Method: Being Male, Seeing Myth and the Analysis of Structures of Gender and Sexuality in the Eastern Interior of Fiji." In *Gendered Fields: Women, Men and Ethnography,* edited by Diane Bell, Pat Caplan, and Wazir Jahan Karim, 63–77. London: Routledge.

Abu-Lughod, Lila. 1993. *Writing Women's Worlds: Bedouin Stories.* Berkeley and Los Angeles: Univ. of California Press.

Allen, Paula Gunn, ed. 1990. *Spider Woman's Granddaughters: Traditional Tales and Contemporary Writing by Native American Women.* London: Women's Press.

Altman, Dennis. 1982. *The Homosexualization of America, the Americanization of the Homosexual.* Boston: Beacon Press.

Anderson, Margaret L. 1992. "From the Editor." *Gender and Society* 6:165–68.

Andrews, Alice. 1989. "Women in Applied Geography." In *Applied Geography: Issues, Questions, and Concerns,* edited by Martin Kenzer, 193–204. Dordrecht: Kluwer Academic Publishers.

Anzaldúa, Gloria, ed. 1990. *Making Face, Making Soul: Haciendo Caras.* San Francisco: Aunt Lute.

Archer, Kevin. 1997. "The Limits to the Imagineered City: Sociospatial Polarization in Orlando." *Economic Geography* 73:322–36.

———. 1996. "Packaging the Place: Development Strategies in Tampa and Orlando, Florida." In *Local Economic Development in Europe and the Americas,* edited by Chistophe Demaziere and Patricia A. Wilson, 239–63. London: Mansell.

————. 1995. "A Folk Guide to Geography as a Holistic Science." *Journal of Geography* 94:404–11.

————. 1994. "Possibilities for Accommodating Dual-Career Faculty Couples at the University of South Florida." Unpublished manuscript, Department of Geography, Univ. of South Florida, Tampa, Fla.

Ashley, Kathleen, Leigh Gilmore, and Gerald Peters, eds. 1994. *Autobiography and Postmodernism.* Amherst: Univ. of Massachusetts Press.

Baisnee, Valerie. 1997. *Gendered Resistance: The Autobiographies of Simone de Beauvoir, Maya Angelou, Janet Frame and Marguerite Duras.* Atlanta: Rodopi.

Barnes, Trevor J. 1996. *Logics of Dislocation: Models, Metaphors, and Meanings of Economic Space.* New York: Guilford.

Barnes, Trevor J., and James Duncan, eds. 1992. *Writing Worlds: Discourse, Text and Metaphor in the Representation of Landscape.* London: Routledge.

Barthes, Roland. 1977. *Roland Barthes by Roland Barthes.* Berkeley and Los Angeles: Univ. of California Press.

Beck, Ulrich. 1992. *Risk Society.* London: Sage.

Behar, Ruth, and Deborah A. Gordon, eds. 1995. *Women Writing Culture.* Berkeley and Los Angeles: Univ. of California Press.

Bell, David, Jon Binnie, Julia Cream, and Gill Valentine. 1994. "All Hyped Up and No Place to Go." *Gender, Place and Culture* 1:31–47.

Bell, David, and Gill Valentine, eds. 1995. *Mapping Desire.* London: Routledge.

Bell, Susan Groag, and Marilyn Yalom, eds. 1990. *Revealing Lives: Autobiography, Biography, Gender.* Albany: State Univ. of New York Press.

Benjamin, Walter. 1978. *Reflections: Essays, Aphorisms, Autobiographical Writings.* New York: Harcourt, Brace, Jovanovich.

————. 1979. *One-Way Street and Other Writings.* London: New Left Books.

Bennett, David. 1984. "Getting There: Experience and Epistemology." Discussion Paper no. 2, Department of Geography. Ottawa: Carleton Univ. Press

Benstock, Shari, ed. 1988. *The Private Self: Theory and Practice of Women's Autobiographical Writings.* London: Routledge.

Bergsten, Karl-Erik. 1988. "Geography: My Inheritance." In *Geographers of Norden: Reflections on Career Experiences,* edited by Anne Buttimer and Torsten Hägerstrand , 61–70. Lund: Lund Univ. Press.

Berman, Mildred. 1977. "Facts and Attitudes on Discrimination as Perceived by AAG Members: Survey Results." *Professional Geographer* 29, no. 1:70–76.

———. 1980. "Millicent Todd Bingham: Human Geographer and Literary Scholar." *Professional Geographer* 32, no. 2:199–204.

———. 1984. "On Being a Woman in American Geography." *Antipode* 16:61–66.

Berry, Brian J. L. 1972. "More on Relevance and Policy Analysis." *Area* 4:77–80.

Bhabha, Homi K. 1983. "Difference, Discrimination and the Discourse of Colonialism." In *The Politics of Theory,* edited by Francis Barker, Peter Hulme, Margaret Iversen, and Diana Loxley, 194–211. Colchester, U.K.: Univ. of Essex.

———. 1984. "Of Mimicry and Man: The Ambivalence of Colonial Discourse." *October* 28:125–33.

———. 1985. "Signs Taken for Wonders: Questions of Ambivalence and Authority under a Tree Outside Delhi, May 1817." *Critical Inquiry* 12:144–65.

Billinge, Mark, Derek Gregory, and Ron L. Martin, eds. 1984. *Recollections of a Revolution.* London: Macmillan.

Blumer, Herbert. 1969. *Symbolic Interactionism.* Englewood Cliffs, N.J.: Prentice-Hall.

Bondi, Liz. 1992. "Gender Symbols in Urban Landscapes." *Progress in Human Geography* 16:157–70.

———. 1999. "Stages on Journeys: Some Remarks about Human Geography and Psychotherapeutic Practice." *Professional Geographer* 51, no. 1:11–24.

Bonnett, Alastair. 1994. "The New Primitives: Identity, Landscape and Cultural Appropriation in the Mythopoetic Men's Movement." *Antipode* 28, no. 3:273–91.

———. 1996. " 'White Studies': The Problems and Projects of a New Research Agenda." *Theory, Culture and Society* 13, no. 2:145–55.

Braidotti, Rosi, Ewa Charkiewizz, Sabine Hausler, and Saskia Wieringa. 1994. *Women, the Environment and Sustainable Development.* London: Zed Books.

Brewster, Anne. 1996. *Reading Aboriginal Women's Autobiography.* Sydney: Sydney Univ. Press.

Brodzki, Bella, and Celeste Schenck, eds. 1988. *Life/Lines: Theorizing Women's Autobiography.* Ithaca, N.Y.: Cornell Univ. Press.

Brossard, Nicole. 1995. "Green Light of Labyrinth Park, La nuit verte du parc Labyrinthe." In *Sexy Bodies: The Strange Carnalities of Feminism,* edited by Elizabeth Grosz and Elspeth Probyn, 128–36. New York: Routledge.

Brown, Anne E., and Marjanne E. Goozé, eds. 1995. *International Women's Writing: New Landscapes of Identity.* Westport, Conn.: Greenwood Press.

Browning, Clyde E., and Barbara Borowiecki, eds. 1982. *Conversations with Geographers: Career Pathways and Research.* Studies in Geography, no. 16. Chapel Hill: Univ. of North Carolina Press.

Browning, Frank. 1996. *A Queer Geography.* New York: Crown.

Bruner, Jerome. 1993. "The Autobiographical Process." In *The Culture of Autobiography: Constructions and Self-Representation,* edited by Robert Folkenflik, 38–56. Stanford, Calif.: Stanford Univ. Press.

Buck-Morss, Susan. 1989. *The Dialectics of Seeing: Walter Benjamin and the Arcades Project.* Cambridge, Mass.: MIT Press.

Buechler, Steve. 1984. "Sex and Class: A Critical Overview of Some Recent Theoretical Work and Some Modest Proposals." *The Insurgent Sociologist* 12:19–32.

Burgess, Jacqueline. 1990. "The Production and Consumption of Environmental Meanings in the Mass Media: A Research Agenda for the 1990s." *Transactions of the Institute of British Geographers* 15:139–61.

Bushong, Allen D. 1984. "Ellen Churchill Semple 1862–1932." *Geographers Biobibliographical Series* 8:87–94.

Butler, Judith. 1990. *Gender Trouble.* London: Routledge.

Buttimer, Anne. 1968. "Social Geography." In MacMillan's *Revised International Encyclopedia of the Social Sciences,* 134–45. New York: Free Press.

———. 1969. "Social Space in Interdisciplinary Perspective." *Geographical Review* LIV, no. 3:417–26.

———. 1971. *Society and Milieu in the French Geographic Tradition.* Association of American Geographers Monograph, no. 6. Chicago: Rand McNally.

———. 1972. "Social Space and the Planning of Residental Areas." *Environment and Behaviour* 4, no. 3:279–318.

———. 1974. *Values in Geography.* Resource Paper no. 24. Washington, D.C.: Commission on College Geography.

———. 1976. "Grasping the Dynamism of the Lifeworld." *Annals of the Association of American Geographers* 66:277–92.

———. 1982. "Musing on Helicon: Root Metaphors in Geography." *Geografiska Annaler* 64B:89–96.

———. 1983a. "Creativity and Context." DIA Paper no. 2. Lund, Sweden: Lund Univ.

———. 1983b. "Perception in Four Keys." In *Environmental Perception and Behaviour: Inventory and Prospect,* edited by Thomas Saarinen and David Seamon, 251–63. Chicago: Univ. of Chicago Press.

———. 1983c. *The Practice of Geography.* London: Longman.

————. 1984a. "Water Symbolism and the Search for Wholeness." In *Vatten bär Livet—Funktioner. Föreställningar och Symbolik,* edited by Reinhold Castensson, 57–92. Linköping: TEMA V Report No. 6. (Abridged edition in *Dwelling, Community and Environment,* edited by David Seamon, 259–80. Dordrecht: Reidel.)

————. 1984b. *Ideal and Wirklichheit in der Angewandten Geographie. Münchner Geographische Hefte,* Nr.51.

————. 1984c. "Meaning-Metaphor-Milieu in the Practice of Geography." In *International Geographical Union Congress Proceedings,* 110–12. Oral presentation to Theme no. 20. Paris: International Geographical Congress.

————. 1985. "Farmers, Fishermen, Gypsies, Guests: Who Identifies?" *Pacific Viewpoint* 26, no. 1:280–344.

————. 1986. "Life Experience as Catalyst for Cross-Disciplinary Communication: Adventures in Dialogue, 1977–1985." DIA Paper no. 3. Lund, Sweden: Lund Univ. Press

————. 1987a. "A Social Topography of Home and Horizon." *Journal of Environmental Psychology* 7:307–19.

————. 1987b. "Life Experience as Catalyst for Cross-Disciplinary Communication." *Journal of Geography and Higher Education* 11, no. 2:133–41.

————. 1987c. "Edgar Kant 1902–1978." *BioBibliographical Studies* 11:71–82.

————. 1988. "Social Implications of Global Change." In *Human Response to Global Change,* edited by Uno Svedin and Bo Heurling, 17–28. Stockholm: FRN Report 88, no. 3.

————. 1989. "Mirrors, Masks and Diverse Milieux." In *Issues of the Behavioural Environment: Essays in Reflection, Applications, and Re-evaluations,* edited by Frederick W. Boal and David N. Livingstone, 253–76. London: Routledge.

————. 1990. "Geography, Humanism, and Global Concern." *Annals of the Association of American Geographers* 80:5–34.

————. 1992a. "Woodland Polyphony." In *Society and the Environment: A Swedish Research Perspective,* edited by Uno Svedin and Britt Hägerhäll Aniansson, 177–98. Dordrecht: Kluwer.

————. 1992b. "Landscape and Life: Appropriate Scales for Sustainable Development." LLASS Working Paper no. 1. Department of Geography, University College Dublin.

————. 1993. *Geography and the Human Spirit.* Baltimore: Johns Hopkins Univ. Press.

————. 1994a. "Edgar Kant and *Heimatkunde:* Balto-Skandia and Regional Identity." In *Geography and National Identity,* edited by David Hooson, 161–83. Oxford: Blackwell.

————. 1994b. "Response to Commentary on 'Grasping the Dynamism of the Lifeworld' (1976)." *Progress in Human Geography* 18, no. 4:501–6.

————. 1994c. "Water Symbolism, Wholeness, and Creativity." *Transforming Art* 4, no. 2:53–62.

————. 1995a. "Gatekeeping Geography Through National Independence: Stories from Harvard and Dublin." *Erdkunde* 49, no. 1:1–16.

————. 1995b. *Landscape and Life: Appropriate Scales for Sustainable Development* [LLASS]. Network research project under DGXII Area 3 European Commission Programme for Environment with partner teams in Germany, Ireland, the Netherlands and Sweden. Final Report on the Project, 130 pp. + Annex A: Team reports from Germany, Ireland, the Netherlands, and Sweden; Annex B: Evaluation of the pluri-disciplinary research forum at UCD and Working Paper series, 20 pp. + 18 working papers.

————. 1995c. " 'Slieveardagh', 1950–1990." Report on the Irish Case Studies in LLASS network project, 214 pp. With Gerry O'Reilly, Taeke Stol, and William Jenkins. Department of Geography, University College Dublin.

————. 1996a. "Reply to Commentaries on 'Values in Geography' (1974)." *Progress in Human Geography* 20, no. 4:513–19.

————. 1996b. "IGU Video Archive/Vidéo-Archives de l'UGI" *IGU Bulletin* 46:75–89.

————. 1996c. "Mapping Mytho-poetic Images of Western Humanism into an Era of Global Environmental Concerns." In *Social Cartography,* edited by Rolland G. Paulston, 141–60. Hamden, Conn.: Garland Publishing.

————. 1996d. "Twilight and Dawn for Geography in Ireland." In *Science, Technology and Medicine in Nineteenth and Twentieth Century Ireland,* edited by Peter J. Bowler, 135–52. Belfast: Institute of Irish Studies.

————. 1996e. "Geography and Humanism in the Late Twentieth Century." In *Companion Encyclopedia of Geography,* edited by Ian Douglas, Richard Huggett, and Mike Robinson, 837–59. London: Routledge.

————. 1996f. "Images de l'Irlande dans les manuels de géographie scolaire." *Espace et Culture* 18:53–74.

————. 1997. "Dublino 1995. Impressioni di Alcuni Giovani Italiani Sull'Irlanda e su gli Irlandesi." *Annali Italiani del Turismo Internazionale* 2, no. 2:31–43.

————. 1998a. "Geography's Stories: Changing States of the Art." *Tijdschrift voor Economishe en Sociale Geografie* 89, no. 1:90–99.

———. 1998b. "Close to Home: Making Sustainability Work at the Local Level." *Environment* 40, no. 3:12–15, 32–40.

———. 1998c. "Landscape and Life: Appropriate Scales for Sustainable Development." *Irish Geography* 31, no. 1:1–33.

Buttimer, Anne, ed. 1992. "Special Issues: History of Geographic Thought." *Geojournal* 26, no. 2.

Buttimer, Anne, John van Buren, and Nancy Hudson-Rodd. 1991. *Land-Life-Lumber-Leisure: Tensions of Local and Global Concern in the Human Use of Woodland*. Interim report on Swedish-Canadian Research Project. Ottawa, Ont.: Royal Society of Canada.

Buttimer, Anne, and Torsten Hägerstrand. 1980. "Invitation to Dialogue." DIA Paper no. 1. Lund, Sweden: Lund Univ.

Buttimer, Anne, and Torsten Hägerstrand, eds. 1988. *Geographers of Norden: Reflections on Career Experiences*. Lund, Sweden: Lund Univ. Press.

Buttimer, Anne, and Anne-Marie McGuaran. 1994. "Impacts of EC Agricultural Politics on the Irish Rural landscape." *Acta Universitatis Carolinae: Geografica* 29, no. 1:5–17.

Buttimer, Anne, and Dennis Graham Pringle. 1996. *Bibliography of Irish Geography 1991–1995*. Dublin: Royal Irish Academy.

Buttimer, Anne, and David Seamon. 1980. *Human Experience of Place and Space*. London: Croom Helm.

Buttimer, Anne, and Taeke Stol. 1997. "Circuits of Calories: Flows of Food and Energy in Germany, Ireland, the Netherlands and Sweden 1960–1990." *European Review* 4, no. 3:1–22.

Butz, David. 1993. "Developing Sustainable Communities: Community Development and Modernity in Shimshal, Pakistan." Ph.D. diss., McMaster University, Hamilton, Ont., Canada.

———. 1995. "Legitimating Porter Regulation in an Indigenous Mountain Community in Northern Pakistan." *Environment and Planning D: Society and Space* 13:381–414.

———. 1996. "Sustaining Indigenous Communities." *Canadian Geographer* 40:36–53.

———. 1997. "Subjectivity and Resistance among Shimshali Porters." Paper presented at the Annual Meeting of the Association of American Geographers, Fort Worth, Texas, 1–5 April.

Butz, David, and John Eyles. 1997. "Reconceptualizing Senses of Place: Social Relations, Ideology and Ecology." *Geografiska Annaler* 79B:1–25.

Castells, Manuel. 1983. *The City and the Grassroots*. Berkeley and Los Angeles: Univ. of California Press.

Caws, Mary Ann. 1990. "Personal Criticism: A Matter of Choice." In *The Women of Bloomsbury*, 1–8. London: Routledge.

Chauncey, George. 1994. *Gay New York: Gender, Urban Culture and the Making of the Gay Male World 1890–1940.* New York: Basic Books.

Chouinard, Vera, Ruth Fincher, and Michael Webber. 1984. "Empirical Research in Scientific Human Geography." *Progress in Human Geography* 8:347–80.

Chouinard, Vera, and Ali Grant. 1995. "On Not Being Anywhere Near the 'Project': Revolutionary Ways of Putting Ourselves in the Picture." *Antipode* 27, no. 2:137–66.

Clark, Gordon. 1981. "Democracy and the Capitalist State: Towards a Critique of the Tiebout Hypothesis." In *Political Studies from Spatial Perspectives: Anglo-American Essays on Political Geography,* edited by Alan D. Burnett and Peter J. Taylor, 111–31. New York: John Wiley and Sons.

Clark, Gordon, and Michael Dear. 1984. *State Apparatus: Structures and Language of Legitimacy.* Boston: Allen and Unwin.

Clifford, James. 1997. *Routes: Travel and Translation in the Late Twentieth Century.* Cambridge, Mass.: Harvard Univ. Press.

Cloke, Paul, Chris Philo, and David Sadler. 1991. *Approaching Human Geography: An Introduction to Contemporary Theoretical Debates.* New York: Guilford.

Coates, Bryan, R. J. Johnston, and Paul L. Knox, 1977. *Geography and Inequality.* London: Oxford Univ. Press.

Coleman, Linda, ed. 1997. *Women's Life-Writing: Finding Voice/Building Community.* Bowling Green, Ohio: Bowling Green State Univ. Popular Press.

Cook, Ian. 1995. "Constructing the Exotic: The Case of Tropical Fruit." In *Geographical Worlds,* edited by John Allen and Doreen Massey, 137–42. Oxford: Oxford Univ. Press.

———. 1997. "A Grumpy Thesis: Geography, Autobiography, Pedagogy." Ph.D. diss., University of Bristol, Bristol, U.K.

———. 1998. " 'You Want to Be Careful You Don't End Up Like Ian. He's All Over the Place': Autobiography in/of an Expanded Field (The Director's Cut)." Research Papers in Geography no. 34. Falmer, U.K.: Univ. of Sussex.

Cornwall, Andrea, and Nancy Lindisfarne, eds. 1993. *Dislocating Masculinity: Comparative Ethnographies.* London: Routledge.

Cornwell, Jocelyn. 1984. *Hard-earned Lives.* London: Tavistock.

Cream, Julia. 1995. "Women on Trial: A Private Pillory?" In *Mapping the Subject: Geographies of Cultural Transformation,* edited by Steve Pile and Nigel Thrift, 158–69. London: Routledge.

Cresswell, Tim. 1998. "The Peninsular of Submerged Hope: Ben Reitman's Social Geography." *Geoforum* 29, no. 2:207–16.

Currie, Dawn, and Hamida Kazi. 1987. "Academic Feminism and the Process of De-radicalization: Re-examining the Issues. *Feminist Review* 25:77–98.

D'Emilio, John. 1981. "Gay Politics, Gay Communities: The San Francisco Experience." *Socialist Review* 55:77–104.

————. 1983. *Sexual Politics, Sexual Communities: The Making of a Homosexual Minority in the United States, 1940–1970.* Chicago: Univ. of Chicago Press.

Davis, Kathy. 1997. "Embody-ing Theory: Beyond Modernist and Postmodernist Readings of the Body." In *Embodied Practices: Feminist Perspectives on the Body,* edited by Kathy Davis, 1–23. London: Sage.

Davis, Mike. 1992. *City of Quartz: Excavating the Future in Los Angeles.* New York: Vintage Books.

Dear, Michael. 1981. "A Theory of the Local State." In *Political Studies from Spatial Perspectives: Anglo-American Essays on Political Geography,* edited by Alan D. Burnett and Peter J. Taylor, 183–200. New York: John Wiley and Sons.

Dear, Michael, and Allen Scott. 1981. "Towards a Framework for Analysis." In *Urbanization and Urban Planning in Capitalist Society,* edited by Michael Dear and Allen J. Scott, 3–16. London and New York: Methuen.

Dear, Michael J., and S. Martin Taylor. 1982. *Not on Our Street.* London: Pion.

Denzin, Norman. 1989. *Interpretive Biography.* London: Sage.

Derrida, Jacques. 1985. *The Ear of the Other: Otobiography, Transference, Translation: Texts and Discussions with Jacques Derrida.* New York: Schocken Books.

Domb, Risa, ed. 1996. *New Women's Writing from Israel.* London: Vallentine Mitchell.

Domosh, Mona. 1997. "With 'Stout Boots and a Stout Heart': Historical Methodology and Feminist Geography." In *Thresholds in Feminist Geography: Difference, Methodology, Representation,* edited by John Paul Jones III, Heidi J. Nast, and Susan M. Roberts, 225–37. Boulder, Colo.: Rowman and Littlefield.

Donovan, Jenny. 1986. *We Don't Buy Sickness, It Just Comes.* Aldershot: Gower.

Doyal, Lesley. 1979. *The Political Economy of Health.* London: Pluto Press.

Duncan, Nancy, ed. 1996. *Body/Space: Destabilizing Geographies of Gender and Sexuality.* New York: Routledge.

Duncan, Simon, and Mark Goodwin. 1982a. "The Local State and Restructuring Social Relations: Theory and Practice." *International Journal of Urban and Regional Research* 6:157–86.

————. 1982b. "The Local State, Functionalism, Autonomy and Class Relations in Cockburn and Saunders." *Political Geography Quarterly* 1:77–96.

Dyck, Isabel. 1989. "Integrating Home and Wage Workplace: Women's Daily Lives in a Canadian Suburb." *Canadian Geographer* 33:329–41.

Eakin, Paul John. 1985. *Fictions in Autobiography: Studies in the Art of Self Invention.* Princeton, N.J.: Princeton Univ. Press.

Ecological Society of America. 1993. "Career Options for Dual-Career Couples: Results of an ESA Survey on Soft Money Research Positions and Shared Positions." *Bulletin of the Ecological Society of America* 74:146–52.

Edel, Matthew. 1981. "Capitalism, Accumulation and the Explanation of Urban Phenomena." In *Urbanization and Urban Planning in Capitalist Society,* edited by Michael Dear and Allen J. Scott, 19–44. London and New York: Methuen.

Elbaz, Robert. 1987. *The Changing Nature of the Self: A Critical Study of the Autobiographical.* Iowa City: Univ. of Iowa Press.

Ellroy, James. 1996. *My Dark Places.* New York: Knopf.

England, Kim V. L. 1994. "Getting Personal: Reflexivity, Positionality, and Feminist Research." *Professional Geographer* 46, no. 1:80–89.

Enloe, Cynthia. 1990. *Bananas, Beaces and Bases: Making Feminist Sense of International Politics.* Berkeley and Los Angeles: Univ. of California Press.

Entrikin, Nick. 1976. "Contemporary Humanism in Geography." *Annals of the Association of American Geographers* 66:615–32.

Escobar, Arturo. 1991. "Anthropology and the Development Encounter: The Making and Marketing of Development Anthropology." *American Ethnologist* 18:658–82.

————. 1995. *Encountering Development: The Making and Unmaking of the Third World.* Princeton, N.J.: Princeton Univ. Press.

Eyles, John. 1971. "Pouring New Sentiments into Old Theories." *Area* 3:242–50.

————. 1974. "Social Theory and Social Geography." *Progress in Geography* 6:27–87.

————. 1985. *Senses of Place.* Warrington: Silverbrook Press.

————. 1987a. "Poverty, Deprivation and Social Planning." In *Social Geography,* edited by Michael Pacione, 201–51. London: Croom Helm.

————. 1987b. "Housing Advertisments as Signs." *Geografiska Annaler* 69B:93–105.

————. 1997. "Environmental Health Research." *Health and Place* 3:1–13.

Eyles, John, ed. 1988. *Research in Human Geography.* Oxford: Blackwell.

Eyles, John, and Jenny Donovan. 1986. *The Social Effects of Health Policy: Ex-*

periences of Health and Health Care in Contemporary Britain. Aldershot, U.K. and Brookfield,Vt.: Gower.

Eyles, John, and Walter G. Peace. 1990. "Signs and Symbols in Hamilton: An Iconology of Steeltown." *Geografiska Annaler* 73B, nos. 2–3:73–88.

Eyles, John, and Kevin J. Woods. 1983. *The Social Geography of Medicine and Health.* London: Croom Helm.

Eyles, John, Stephen Birch, Jerry Hurley, Brian Hutchison, and Shelley Chambers. 1991. "A Needs-based Methodology for Allocating Health-care Resources in Ontario." *Social Science and Medicine* 33:489–500.

Eyles, John, Doug Sider, Jamie Baxter, Martin Taylor, and Dennis Willms. 1993. "The Social Construction of Risk in a Rural Community." *Risk Analysis* 13:281–90.

Eyles, John, Martin Taylor, Doug Sider, and Nancy Johnson. 1993. "Worrying about Waste." *Social Science and Medicine* 37:805–12.

Eyles, John, Stephen Birch, and K. Bruce Newbold. 1995. "Delivering the Goods?" *Journal of Health and Social Behavior* 36:322–32.

Eyles, John, and David M. Smith, eds. 1988. *Qualitative Methods in Human Geography.* Cambridge: Polity Press.

Fellows, Will. 1996. *Farm Boys: Lives of Gay Men from the Rural Midwest.* Madison: Univ. of Wisconsin Press.

Fincher, Ruth. 1981. "Local Implementation Strategies in the Urban Built Environment." *Environment and Planning A* 13:1233–52.

———. 1983. "The Inconsistency of Eclecticism." *Environment and Planning A* 15:607–22.

———. 1984. "The State Apparatus and the Commodification of Quebec's Housing Cooperatives." *Political Geography Quarterly* 3:127–43.

Fincher, Ruth, and Sue Ruddick. 1983. "Transformation Possibilities Within the Capitalist State: Cooperative Housing and Decentralized Health Care in Quebec." *International Journal of Urban and Regional Research* 7:44–71.

Fiske, Jonathan. 1987. "Ethnosemiotics: Some Personal and Theoretical Reflections." *Cultural Studies* 4:85–99.

Fleming, Marie. 1988. *The Geography of Freedom: The Odyssey of Elisée Reclus.* Montréal: Black Rose Books.

Forbes, Jean, ed. 1973. *Studies in Social Science and Planning.* Edinburgh and London: Scottish Academic Press.

Foucault, Michel. 1967. *Madness and Civilization.* London: Tavistock.

———. 1972. *The Archaeology of Knowledge and the Discourse on Language.* New York: Pantheon.

———. 1980. *The History of Sexuality, Vol. 1.* New York: Vintage Books.

Fowler, Lois J., and David H. Fowler, eds. 1990. *Revelations of Self: American Women in Autobiography.* Albany: State Univ. of New York Press.

Friedman, Susan Stanford. 1988. "Women's Autobiographical Selves: Theory and Practice." In *The Private Self: Theory and Practice of Women's Autobiographical Writings,* edited by Shari Benstock, 34–62. Chapel Hill: Univ. of North Carolina Press.

Frisby, David. 1985. *Fragments of Modernity: Theories of Modernity in the Work of Simmel, Kracauer, and Benjamin.* London: Polity Press.

Fuss, Diana. 1989. *Essentially Speaking: Feminism, Nature, and Difference.* New York: Routledge.

Gaarder, Jostein. 1996. *Sophie's World: A Novel about the History of Philosophy.* Translated by Paula Møller, 1994. Reprint. New York: Berkley Books.

Gamson, William A., and Andre Modigliari. 1989. "Media Discourse and Public Opinion on Nuclear Power." *American Journal of Sociology* 95:1–37.

Garvin, Theresa, and John Eyles. 1997. "The Sun Safety Metanarrative." *Policy Sciences* 30:47–70.

Geertz, Clifford. 1973. *The Interpretation of Cultures.* New York: Basic Books.

———. 1995. *After the Fact: Two Countries, Four Decades, One Anthropologist.* Cambridge, Mass.: Harvard Univ. Press.

Gendzier, Irene. 1985. *Managing Political Change: Social Scientists and the Third World.* Boulder, Colo.: Westview Press.

Gibson-Graham, J.-K. 1994. " 'Stuffed If I Know': Reflections on Postmodern Feminist Social Research." *Gender, Place and Culture* 1:205–24.

———. 1999. "Queer(y)ing Capitalism In and Out of the Classroom." *Journal of Geography in Higher Education* 23, no. 1:80–85.

Giddens, Anthony. 1973. *Class Structure of Advanced Societies.* London: Macmillan.

———. 1976. *New Rules of Sociological Method.* London: Hutchinson.

———. 1979. *Central Problems in Social Theory.* Berkeley and Los Angeles: Univ. of California Press.

———. 1981. *A Contemporary Critique of Historical Materialism,* Vol. 1. Berkeley and Los Angeles: Univ. of California Press.

———. 1991. *Modernity and Self-identity.* Cambridge: Polity Press.

Gilbert, Melissa. 1994. "The Politics of Location: Doing Feminist Research at 'Home'." *Professional Geographer* 46, no. 1:90–96.

Gilmore, Leigh. 1994. *Autobiographies: A Feminist Theory of Women's Self-Representation.* Ithaca, N.Y.: Cornell Univ. Press.

Gilroy, Paul. 1993. *The Black Atlantic: Modernity and Double Consciousness.* London and New York: Verso.

Giroux, Henry A. 1997. *Pedagogy and the Politics of Hope: Theory, Culture, and Schooling: A Critical Reader.* Boulder: Westview Press.

Goodman, Katherine. 1986. *Dis-Closures: Women's Autobiography in Germany Between 1790 and 1914.* Bern: Lang.

Gornick, Vivian. 1983. *Women in Science: Portraits from a World in Transition.* New York: Simon and Schuster.

Gould, Peter. 1988. "Expose Yourself to Geographic Research." In *Research in Human Geography,* edited by John Eyles, 11–27. Oxford: Blackwell.

Gould, Peter, and Gunnar Olsson. 1982. *A Search for Common Ground.* London: Pion.

Greene, Gail, and Coppélia Kahn, eds. 1993. *Changing Subjects: The Making of Feminist Literary Criticism.* London: Routledge.

Gregory, Derek. 1994. *Geographical Imaginations.* Oxford: Blackwell.

Gregory, Derek, and John Urry, eds. 1985. *Social Relations and Spatial Structures.* New York: St. Martin's Press.

Grosz, Elizabeth. 1992. "Bodies-Cities." In *Sexuality and Space,* edited by Beatrice Colomina, 241–53. New York: Princeton Architectural Press.

———. 1995. *Space, Time and Perversion.* New York: Routledge.

Grosz, Elizabeth, and Elspeth Probyn, eds. 1995. *Sexy Bodies: The Strange Carnalities of Feminism.* New York: Routledge.

Habermas, Jürgen. 1981. *The Theory of Communicative Action: Reason and the Rationalization of Society.* Boston: Beacon Press.

———. 1987. *The Philosophical Discourse on Modernity.* Cambridge, Mass.: MIT Press.

Hägerstrand, Torsten. 1970. "What about People in Regional Science?" *Papers of the Regional Science Association* 24:7–21.

Hannigan, John Andrew. 1995. *Environmental Sociology.* London: Routledge.

Hanson, Susan. 1997. "As the World Turns: New Horizons in Feminist Geographic Methodologies." In *Thresholds in Feminist Geography: Difference, Methodology, Representation,* edited by John Paul Jones III, Heidi J. Nast, and Susan M. Roberts, 119–28. Boulder, Colo.: Rowman and Littlefield.

Haraway, Donna. 1988. "Situated Knowledges: The Science Question in Feminism and the Privilege of Partial Perspective." *Feminist Studies* 14, no. 3:575–99.

Hart, John Fraser. 1979. "Postwar Years." *Annals of the Association of American Geographers* 69:109–114.

Hart, Lynda, and Peggy Phelan, eds. 1993. *Acting Out: Feminist Performances.* Ann Arbor: Univ. of Michigan Press.

Harvey, David. 1971. "Social Processes, Spatial Form and the Redistribution of Real Income in an Urban System." In *Regional Forecasting,* edited by Michael Chisholm, Allen E. Frey, and Peter Haggett, 267–300. London: Butterworth.

———. 1972. "Revolutionary and Counterrevolutionary Theory in Geography and the Problem of Ghetto Formation." *Antipode* 4, no. 2:1–13.

———. 1973. *Social Justice and the City.* Baltimore: Johns Hopkins Univ. Press.

———. 1974. "What Kind of Geography for which Kind of Public Policy?" *Transactions of the Institute of British Geographers* 63:18–24.

———. 1985. *The Urbanization of Capital.* Baltimore: Johns Hopkins Univ. Press.

———. 1989. *The Condition of Postmodernity.* Cambridge, Mass.: Blackwell.

———. 1990. "From Space to Place and Back Again: Reflections on the Condition of Postmodernity." Presented at Symposium on FUTURES, the Tate Gallery, London, November 24.

———. 1996. *Justice, Nature and the Geography of Difference.* Cambridge, Mass.: Blackwell.

Heilbrun, Carolyn. 1988. *Writing a Woman's Life.* London: Women's Press.

Herndl, Diane Price. 1993. *Invalid Women: Figuring Feminine Illness in American Fiction and Culture, 1840–1940.* Chapel Hill: Univ. of North Carolina Press.

Hill, Christopher. 1971. *Intellectual Origins of the English Revolution.* Oxford: Clarendon Press.

———. 1990. *The World Turned Upside Down.* Harmondsworth, Eng.: Penguin.

Hiroko, Tomida. 1996. *Japanese Writing on Women's History.* Oxford: Nissan Institute of Japanese Studies.

Holmes, Diana. 1996. *French Women's Writing: 1848–1994.* London: Athlone.

Holmstrom, N. 1981. " 'Women's Work', the Family, and Capitalism." *Science and Society* 45:186–211.

hooks, bell. 1992. *Black Looks: Race and Representation.* Boston, Mass.: South End Press.

———. 1994. *Teaching to Transgress: Education in the Practice of Freedom.* New York: Routledge.

———. 1997. *Bone Black: Memories of Girlhood.* London: Women's Press.

Horgan, John. 1996. *The End of Science.* New York: Broadway Books.

Horst, Oscar H. 1981. "Lucia Harrison: In Recognition of Her Contributions to the Role of Women in Geography." In *Papers in Geography in Honor of Lucia C. Harrison,* edited by Oscar H. Horst. Muncie, Ind.: Conference of Latin Americanist Geographers.

Hoskins, William George. 1955. *The Making of the English Landscape*. London: Hodder and Stoughton.

Humlum, Johannes. 1988. "Peregrinations and Irrigation." In *Geographers of Norden: Reflections on Career Experiences,* edited by Anne Buttimer and Torsten Hägerstrand, 164–89. Lund: Lund Univ. Press.

Iannantuono, Adele, and John Eyles. 1997. "Meanings in Policy." *Social Science and Medicine* 44:1611–21.

Ibarra, Peter R., and John I. Kitsuse. 1993. "Vernacular Constituents of Moral Discourse." In *Reconsidering Social Constructionism,* edited by James A. Holstein and Gales Miller, 25–58. New York: Aldine de Gruyter.

Ingold, Thomas. 1992. "Culture and Perception of the Environment." In *Bush Base, Forest Farm: Culture, Environment and Development,* edited by Elizabeth Croll and David Parkin, 39–56. New York: Routledge.

Jackson, Jean E. 1990. " 'I Am a Fieldnote': Fieldnotes as a Symbol of Professional Identity." In *Fieldnotes: The Making of Anthropology,* edited by Roger Sanjek, 3–33. Ithaca, N.Y.: Cornell Univ. Press.

Jackson, Peter. 1989. *Maps of Meaning: An Introduction to Cultural Geography.* London: Unwin Hyman.

Jackson, Peter, and Susan Smith. 1984. *Exploring Social Geography.* London: Allen and Unwin.

James, Adeola, ed. 1990. *In Their Own Voices: African Women Writers Talk.* London: Currey.

James, Preston Everitt. 1981. *All Possible Worlds: A Historical Geographical Union.* New York: Wiley.

Jelenik, Estelle C., ed. 1980. *Women's Autobiographies: Essays in Criticism.* Bloomington: Indiana Univ. Press.

Johnston, R. J. 1979. *Geography and Geographers.* London: Arnold.

———. 1984. "A Foundling Floundering—World Three." In *Recollections of a Revolution,* edited by Michael Billinge, Derek Gregory, and Ron L. Martin, 39–56. London: Macmillan.

———. 1991. *Geography and Geographers: Anglo-American Human Geography since 1945.* 4th ed. New York: Edward Arnold.

Jones, John Paul III, and Pamela Moss. 1995. "Democracy, Identity, and Space." *Environment and Planning D: Society and Space* 13, no. 3:253–57.

Jouve, Nicole Ward. 1991. *White Woman Speaks with Forked Tongue: Criticism as Autobiography.* London: Routledge.

Kane, Jock, and Betty Kane. 1994. *'No Wonder We Were Rebels': The Kane Story.* Doncaster, U.K.: Askew Design and Print.

Kass-Simon, Gabriele, and Patricia Farnes. 1990. *Women of Science: Righting the Record.* Bloomington: Indiana Univ. Press.

Katz, Cindi. 1992. "All the World Is Staged: Intellectuals and the Projects of Ethnography." *Environment and Planning D: Society and Space* 10:495–510.

———. 1994. "Playing the Field: Questions of Fieldwork in Geography." *Professional Geographer* 41, no. 1:67–72.

Katz, Cindi, and Andrew Kirby. 1991. "In the Nature of Things: The Environment and Everyday Life." *Transactions of the Institute of British Geographers* 16:259–71.

Katz, Cindi, and Janice Monk. 1993. *Full Circles: Geographies of Women over the Life Course.* London: Routledge.

Keat, Russel, and John Urry. 1982. *Social Theory as Science.* 2nd ed. London: Routledge and Kegan Paul.

Keith, Michael, and Steve Pile, eds. 1993. *Place and the Politics of Identity.* London: Routledge.

Kenzer, Martin. 1998. "Honeymoon in New England: A Fresh Look at Carl Sauer's Year in Salem, MA." Paper presented at the Annual Meeting of the Association of American Geographers, Boston, March.

Ketteringham, William. 1979. "Gay Public Space and the Urban Landscape: A Preliminary Assessment." Presented at the Annual Meeting of the Association of American Geographers, Philadelphia, Penn., April.

———. 1983. "The Broadway Corridor: Gay Businesses as Agents of Revitalization in Long Beach, California." Presented at the Annual Meeting of the Association of American Geographers, Denver, Colo., April.

Knopp, Lawrence. 1994. "Social Justice, *Sexuality* and the City." *Urban Geography* 15:644–60.

———. 1996. "Critical Geography Meets Queer Studies: Political Economies of Space and Cultural Politics of Sexuality." Williamson Bequest Lecture Series, Univ. of North Carolina, Chapel Hill, N. C., October.

———. 1997. "Rings, Circles and Perverted Justice: Gay Judges and Moral Panic in Contemporary Scotland." In *Geographies of Resistance,* edited by Michael Keith and Steve Pile, 168–83. London and New York: Routledge.

———. 1998. "Sexuality and Urban Space: Gay Male Identity Politics in the United States, the United Kingdom, and Australia." In *Cities of Difference,* edited by Ruth Fincher and Jane Jacobs, 149–176. New York: Guilford.

Kobayashi, Audrey. 1994. "Colouring the Field: Gender, 'Race', and the Politics of Fieldwork." *Professional Geographer* 46:73–80.

———. 1997. "The Paradox of Difference and Diversity (or, Why the

Threshold Keeps Moving)." In *Thresholds in Feminist Geography: Difference, Methodology, Representation,* edited by John Paul Jones III, Heidi J. Nast, and Susan M. Roberts, 3–9. Boulder, Colo.: Rowman and Littlefield.

Kobayashi, Audrey, and Linda Peake. 1994. "Unnatural Discourse: 'Race' and Gender in Geography. *Gender, Place and Culture* 1:225–43.

Kobayashi, Audrey, Linda Peake, Hal Benenson, and Katie Pickles. 1994. "Placing Women and Work." In *Women, Work, and Place,* edited by Audrey Kobayashi, xi–xlv. Montréal and Kingston: McGill-Queen's Univ. Press.

Korten, David. 1995. *When Corporations Rule the World.* San Francisco: Kumarian Press.

Ladd, Brian. 1997. *The Ghosts of Berlin.* Chicago: Univ. of Chicago Press.

Lake, Marilyn. 1994. "Between Old World 'Barbarism' and Stone Age 'Primitivism': The Double Difference of the White Australian Feminist." In *Australian Women: Contemporary Feminist Thought,* edited by Norma Grieve and Alisa Burns, 80–91. Oxford: Oxford Univ. Press.

Lamont, Marc. 1984. "How to Become a Dominant French Philosopher." *American Journal of Sociology* 93:584–622.

Lauria, Mickey. 1986. "Toward a Specification of the Local State: State Intervention Strategies in Response to a Manufacturing Plant Closure." *Antipode* 18:39–65.

Lauria, Mickey, and Lawrence M. Knopp. 1985. "Toward an Analysis of the Role of Gay Communities in the Urban Renaissance." *Urban Geography* 6:152–169.

Lazarus, Richard S., and Susan Folkman. 1984. *Stress, Appraisal and Copying.* New York: Springer.

Lejeune, Philippe, ed. 1988. *On Autobiography.* Minneapolis, Minn.: Univ. of Minnesota Press.

Levine, Martin. 1979. "Gay Ghetto." *Journal of Homosexuality* 4:363–77.

Lewis, Jane. 1984. "The Role of Female Employment in the Industrial Restructuring and Regional Redevelopment of the U.K." *Antipode* 16:47–60.

Ley, David. 1974. *The Black Inner City as Frontier Outpost.* Washington, D.C.: Association of American Geography.

———. 1977. "Social Geography and the Taken-for-granted World." *Transactions of the Institute of British Geographers* 2:498–512.

Lionnet, Françoise. 1988. "*Métissage,* Emancipation, and Female Textuality in Two Francophone Writers." In *Life/Lines: Theorizing Women's Autobiography,* edited by Bella Brodzki and Celeste Schenck, 260–78. Ithaca, N.Y.: Cornell Univ. Press.

Litva, Andrea, and John Eyles. 1994. "Health or Healthy?" *Social Science and Medicine* 39:1083–91.

———. 1995. "Coming Out." *Health and Place* 1:5–14.

Livingstone, David N. 1992. *The Geographical Tradition.* Oxford: Blackwell.

Lowenthal, David, ed. 1967. *Environmental Perception and Behaviour.* Research Paper no. 9. Univ. of Chicago: Department of Geography Research Series.

Lubchenco, Jane, and Bruce A. Menge. 1993. "Split Positions Can Provide a 'Sane Track': A Personal Account." *Bioscience* 43:243–48.

Lunn, Eugene. 1984. *Marxism and Modernism: An Historical Study of Lukács, Brecht, Benjamin and Adorno.* Berkeley and Los Angeles: Univ. of California Press.

MacCole, John. 1993. *Walter Benjamin and the Antinomies of Tradition.* Ithaca, N.Y.: Cornell Univ. Press.

MacDonald, Kenneth, and David Butz. 1998. "Investigating Portering Relations as a Locus for Transcultural Interaction in the Karakoram Region of Northern Pakistan." *Mountain Research and Development* 18, no. 4:333–43.

Mackenzie, Suzanne, and Damaris Rose. 1983. "Industrial Change, the Domestic Economy and Home Life." In *Redundant Spaces in Cities and Regions?* edited by James Anderson, Simon Duncan, and Ray Hudson, 155–200. London: Academic Press.

Madge, Clare. 1993. "Conference Rantings 1993." *Women and Geography Study Group Newsletter* 2:7–8.

Malinowski, Bronislaw. 1922. *Argonauts of the Western Pacific: An Account of Native Enterprise and Adventure in the Archipelagoes of Melanesian New Guinea.* New York: G. Routledge and Sons.

Marcus, George. 1995. "Ethnography in/of the World System." *Annual Review of Anthropology* 24:95–117

Marshall, T. H. 1965. *Class, Citizenship and Social Development.* New York: Doubleday Anchor.

Massey, Doreen. 1991a. "A Global Sense of Place." *Marxism Today* June:24–29.

———. 1991b. "Flexible Sexism." *Environment and Planning D: Society and Space* 9:31–57.

———. 1995. *Space, Place and Gender.* Cambridge: Polity.

Mattingly, Doreen, and Karen Falconer Al-Hindi. 1995. "Should Women Count? A Context for the Debate." *Professional Geographer* 47, no. 4:427–35.

McDowell, Linda. 1983. "Towards an Understanding of the Gender Division of Urban Space." *Environment and Planning D: Society and Space* 1:59–72.

McDowell, Linda. 1992a. "Doing Gender: Feminism, Feminists and Research Methods in Human Geography." *Transactions, Institute of British Geographers* 17:399–416.

———. 1992b. "Multiple Voices: Speaking from Inside and Outside 'The Project'." *Antipode* 24, no. 1:56–71.

McGee, Terry G. 1995. "Eurocentrism and Geography: Reflections on Asian Urbanization." In *Power of Development,* edited by Jonathan Crush, 192–207. New York: Routledge.

McNee, Robert. 1984. "If You Are Squeamish." *East Lakes Geographer* 19:16–27.

———. 1985. "Will Gays Find Justice in the Queen City?" *Urban Resources* 2: C1-C5.

Mead, George H. 1934. *Mind, Self and Society.* Chicago: Univ. of Chicago Press.

Mead, Margaret. 1928. *Coming of Age in Samoa.* New York: W. Morrow.

Mendlovitz, Saul. H. 1998. "Statement of the International Peoples' Tribunal on Human Rights and the Environment: Sustainable Development in the Context of Globalization." *Alternatives* 23:109–46.

Miller, Nancy K. 1991. *Getting Personal: Feminist Occasions and Other Autobiographical Acts.* New York: Routledge.

Mohanty, Chandra Talpade. 1994. "Under Western Eyes: Feminist Scholarship and Colonial Discourses." In *Colonial Discourses and Post-Colonial Theory: A Reader,* edited by Patrick Williams and Laura Chrisman, 196–221. Hertfordshire, U.K.: Harvester Wheatsheaf.

Mohanty, Chandra Talpade, Ann Russo, and Lourdes Torres, eds. 1991. *Third World Women and the Politics of Feminism.* Bloomington: Indiana Univ. Press.

Monk, Janice. 1998. " 'The Women Were Always Welcome at Clark.' " *Economic Geography* extra issue:14–30.

———. 1997. "Marginal Notes on Representation." In *Thresholds in Feminist Geography: Difference, Methodology, Representation,* edited by John Paul Jones III, Heidi J. Nast, and Susan M. Roberts, 241–53. Boulder, Colo.: Rowman and Littlefield.

Morrill, Richard L. 1984. "Recollections of the Quantitative Revolution's Early Years: The University of Washington 1955–65." In *Recollections of a Revolution: Geography as Spatial Science,* edited by Mark Billinge, Derek Gregory, and Ron Martin, 57–72. London: Macmillan.

Moss, Pamela. 1993. "Production and Reproduction in Waged Domestic

Labour Processes in Housekeeping Services Franchises in Southern Ontario." Chapter 1. Ph.D. diss., McMaster University, Hamilton, Ont., Canada.

———. 1995a. "Inscribing Workplaces: The Social Being in the Spatiality of a Production Process." *Growth and Change: A Journal of Urban Regional Policy* 26 (winter):23–57.

———. 1995b. "Embeddedness in Practice, Numbers in Context." *Professional Geographer* 47, no. 3:442–49.

———. 1995c. "Reflections on the 'Gap' as Part of the Politics of Research Design." *Antipode* 27, no. 1:82–90.

———. 1997a. "Negotiating Spaces in Home Environments: Older Women Living with Arthritis." *Social Science and Medicine* 45:23–33.

———. 1997b. "Spaces of Resistance, Spaces of Respite: Franchise Housekeepers Keeping House in the Workplace and at Home." *Gender, Place and Culture* 4, no. 2:179–96.

———. 1997c. "Researching Chronic Illness: Some Autobiographical Notes." Paper presented at the Annual Meeting of the Institute of British Geographers, Exeter, U.K., January.

———. 1999. "Autobiographical Notes on Chronic Illness." In *Mind and Body Spaces,* edited by Hester Parr and Ruth Butler, 155–166. London: Routledge.

Mouffe, Chantal. 1995. "Post-Marxism, Democracy and Identity." *Environment and Planning D: Society and Space* 13, no. 3:259–65.

Nagar, Richa. 1997. "Exploring Methodological Borderlands Through Oral Narratives." In *Thresholds in Feminist Geography: Difference, Methodology, Representation,* edited by John Paul Jones III, Heidi J. Nast, and Susan M. Roberts, 203–24. Boulder, Colo.: Rowman and Littlefield.

Nägele, Rainer. 1991. *Theater, Theory, Speculation: Walter Benjamin and the Scenes of Modernity.* Baltimore: Johns Hopkins Univ. Press.

Nast, Heidi J. 1994. "Opening Remarks on 'Women in the Field'." *Professional Geographer* 46, no. 1:54–66.

Native Women's Writing Circle. 1996. *Into the Moon: Heart, Mind, Body, Soul.* Toronto: SisterVision Press.

Navarro, Vincente. 1978. *Class Struggle, the State and Medicine.* Edinburgh: Martin Robertson.

Neuman, Shirley, and Smaro Kamboureli, eds. 1986. *Amazing Spaces: Writing Canadian Women Writing.* Edmonton: Longspoon.

Norwood, Vera, and Janice Monk, eds. 1987. *The Desert Is No Lady: Southwestern Landscapes in Women's Writing and Art.* New Haven, Conn.: Yale Univ. Press.

Oberhauser, Ann M. 1995. "Gender and Household Economic Strategies in Rural Appalachia." *Gender, Place and Culture* 2, no. 1:51–70.

Okely, Judith. 1992. "Anthropology and Autobiography: Participatory Experience and Embodied Knowledge." In *Anthropology and Autobiography*, edited by Judith Okely and Helen Callaway, 1–28. London: Routledge.

Okely, Judith, and Helen Callaway, eds. 1992. *Anthropology and Autobiography.* New York: Routledge.

Olney, James, ed. 1980. *Autobiography: Essays Theoretical and Critical.* Princeton, N.J.: Princeton Univ. Press.

———, ed. 1988. *Studies in Autobiography.* New York: Oxford Univ. Press.

Pahl, Ray. 1970. *Patterns of Urban Life.* London: Longman.

———. 1971. *Whose City?* London: Longman.

Paris, Chris. 1983. "The Myth of Urban Politics." *Environment and Planning D: Society and Space* 1:89–108.

Parkin, Frank. 1979. *Class Inequality and Political Order.* London: Paladin.

Paterson, John L. 1984. *David Harvey's Geography.* Totawa, N.J.: Barnes and Noble.

Paules, Greta Foff. 1991. *'Dishing It Out': Power and Resistance in a New Jersey Restaurant.* Philadelphia: Temple Univ. Press.

Peet, Richard. 1977. "The Development of Radical Geography in the United States." *Progress in Human Geography* 1:240–63.

Perrow, Charles. 1984. *Normal Accidents.* New York: Basic Books.

Personal Narratives Group. 1989. *Interpreting Women's Lives: Feminist Theory and Personal Narratives.* Bloomington: Indiana Univ. Press.

Pesman, Ros. 1996. *Duty Free: Australian Women Abroad.* Melbourne and New York: Oxford Univ. Press.

Pfeil, Fred. 1995. *White Guys: Studies in Postmodern Domination and Difference.* London: Verso.

Pickvance, C. G., ed. 1976. *Urban Sociology.* London: Heinemann.

Pile, Steve. 1996. *The Body and the City: Psychoanalysis, Space and Subjectivity.* London: Routledge.

Pile, Steve, and Nigel Thrift, eds. 1995. *Mapping the Subject : Geographies of Cultural Transformation.* London: Routledge.

Podlesney, Eva. 1991. "Blondes." In *The Hysterical Male: New Feminist Theory,* edited by Arthur and Marilouise Kroker, 69–90. Montréal: New World Perspectives.

Porteous, J. Douglas. 1989. *Planned to Death: The Annihilation of a Place Called Howdendyke.* Toronto: Univ. of Toronto Press.

Poster, Mark. 1989. *Critical Theory and Poststructuralism: In Search of a Context.* Ithaca, N.Y.: Cornell Univ. Press.

Powdermaker, Hortense. 1966. *Stranger and Friend.* New York: Norton.

Pratt, Mary Louise. 1992. *Imperial Eyes: Travel Writing and Transculturation.* London: Routledge.

Pulido, Laura. 1997. "Community, Place, and Identity." In *Thresholds in Feminist Geography: Difference, Methodology, Representation,* edited by John Paul Jones III, Heidi J. Nast, and Susan M. Roberts, 11–28. Boulder, Colo.: Rowman and Littlefield.

Rabinow, Paul. 1977. *Reflections of Fieldwork in Morocco.* Berkeley and Los Angeles: Univ. of California Press.

Rex, John, and Robert Moore. 1967. *Race, Community and Conflict.* Oxford: Oxford Univ. Press.

Reynolds, David R. 1981. "The Geography of Social Choice." In *Political Studies from Spatial Perspectives: Anglo-American Essays on Political Geography,* edited by Alan D. Burnett and Peter J. Taylor, 91–109. New York: John Wiley and Sons.

Robinson, Jennifer. 1994. "White Women Researching/Representing 'Others': From Antiapartheid to Postcolonialism?" In *Writing Women and Space: Colonial and Postcolonial Geographies,* edited by Allison Blunt and Gillian Rose, 197–226. New York: Guilford Press.

Rocheleau, Dianne. 1995. "Maps, Numbers, Text and Context: Mixing Methods in Feminist Political Ecology." *Professional Geographer* 47, no. 4:458–66.

Rock, Paul. 1979. *The Making of Symbolic Interactionism.* London: Macmillan.

Rose, Damaris. 1984. "Rethinking Gentrification: Beyond the Uneven Development of Marxist Thinking." *Environment and Planning D: Society and Space* 1:47–74.

Rose, Gillian. 1993. *Feminism and Geography: The Limits to Geographical Knowledge.* Minneapolis: Univ. of Minnesota Press.

———. 1997. "Situating Knowledges: Positionality, Reflexivities and Other Tactics." *Progress in Human Geography* 21, no. 3:305–20.

Rossiter, Margaret W. 1995. *Women Scientists in America: Before Affirmative Action, 1940–1972.* Baltimore: Johns Hopkins Univ. Press.

Routledge, Paul. 1996. "The Third Space as Critical Engagement." *Antipode* 28, no. 4:399–419.

Rubin, Barbara. 1979. "Women in Geography Revisited: Present Status, New Options." *Professional Geographer* 31, no. 2:125–34.

Saltmarsh, Rachel. 1995. "Pit Closure: The Destruction of a Culture." Honors Undergraduate diss., University of Wales, Lampeter, U.K.

———. 1997. Field Notes, Nov.

Sanders, Valerie, ed. 1989. *The Private Lives of Victorian Women: Autobiography in Nineteenth Century England.* New York: Harvester Wheatsheaf.

Sanderson, Marie. 1974. "Mary Somerville: Her Work in Physical Geography." *Geographical Review* 64:410–20.

Sanjek, Roger, ed. 1990. *Fieldnotes: The Making of Anthropology.* Ithaca, N.Y.: Cornell Univ. Press.

Saunders, Peter. 1981. *Social Theory and the Urban Question.* New York: Holmes and Meier.

Sayer, Andrew. 1982. *Method in Social Science: A Realist Approach.* London: Hutchinson.

Schreiner, Barbara, ed. 1992. *A Snake with Ice Water: Prison Writings by South African Women.* Johannesburg, South Africa: COSAW.

Schütz, Alfred. 1972. *The Phenomenology of the Social World.* London: Heinemann.

———. 1973. *Structures of the Lifeworld.* Evanston, Ill.: Northwestern Univ. Press.

Scott, James C. 1990. *Hidden Transcripts: Domination and the Arts of Resistance.* New Haven, Conn.: Yale Univ. Press.

Scott, Joan W. 1991. "The Evidence of Experience." *Critical Inquiry* 17, no. 4:773–97.

Sedgwick, Eve. 1990. *Epistemology of the Closet.* Berkeley and Los Angeles: Univ. of California Press.

Seidler, Victor J. 1995. "Men, Heterosexualities and Emotional Life." In *Mapping the Subject: Geographies of Cultural Transformation,* edited by Steve Pile and Nigel Thrift, 170–91. London: Routledge.

Semple, Ellen Churchill. 1931. *The Geography of the Mediterranean World: Its Relation to Ancient History.* New York: Holt.

Sen, Gita, and Caren Grown. 1987. *Development, Crises, and Alternative Visions.* New York: Monthly Review Press.

Shrethsa, Nanda. 1995. "Becoming a Development Category." In *Power of Development,* edited by Jonathan Crush, 266–77. New York: Routledge.

Sidaway, James. 1997. "The Production of British Geography." *Transactions of the Institute of British Geographers* 22, no. 4:488–504.

Silverman, David. 1972. "Some Neglected Questions about Social Reality." In *New Directions in Sociological Theory,* edited by Paul Filmer. London: Collier-Macmillan.

Sizoo, Edith, ed. 1997. *Women's Lifeworlds: Women's Narratives on Shaping Their Realities.* New York: Routledge.

Smart, Mollie S., and Russell C. Smart. 1990. "Paired Prospects: Dual-Career Couples on Campus." *Academe* 76:33–7.

Smith, Barbara E. 1981. Black Lung: The Social Production of Disease." *International Journal of Health Services* 11:343-59.

Smith, David M. 1973. *The Geography of Well-being in the United States.* New York: McGraw Hill.

———. 1977. *Human Geography.* London: Arnold.

Smith, Neil. 1979. "Toward a Theory of Gentrification: A Back-to-the-City Movement by Capital Not People." *Journal of the American Planning Association* 45:538–48.

———. 1983. "Gentrification and Uneven Development." *Economic Geography* 58:139–55.

Smith, Neil. Forthcoming. *The Geographical Pivot of History: Isaiah Bowman and the Geography of the American Century.* Baltimore: Johns Hopkins Univ. Press.

Smith, Neil, and Michele LeFaivre, 1984. "A Class Analysis of Gentrification." In *Gentrification, Displacement and Neighborhood Revitalization,* edited by J. John Palen and Bruce London, 43–63. Albany: State Univ. of New York Press.

Smith, Sidonie, and Julia Watson, eds. 1992. *De/colonizing the Subject: The Politics of Gender in Women's Autobiography.* Minneapolis: Univ. of Minnesota Press.

Snitow, Ann. 1990. "A Gender Diary." In *Conflicts in Feminism,* edited by Marianne Hirsch and Evelyn Fox, 9–43. London: Routledge.

Snitow, Ann, Christine Stansell, and Sharon Thompson, eds. 1983. *Powers of Desire: The Politics of Sexuality.* New York: Monthly Review Press.

Sohn-Rethel, A. 1978. *Intellectual and Manual Labour: A Critique of Epistemology.* Atlantic Highlands: Humanities Press.

Soja, Edward W. 1989. *Postmodern Geographies: The Reassertion of Space in Critical Social Theory.* London: Verso.

———. 1996. *Thirdspace: Journeys to Los Angeles and Other Real-and-Imagined Places.* Cambridge, Mass.: Blackwell.

Somme, Axel. 1988. "A Geographer of Commitment. An Interview" (by Torsten Hägerstrand). In *Geographers of Norden: Reflections on Career Experiences,* edited by Anne Buttimer and Torsten Hägerstrand , 61–70. Lund: Lund Univ. Press.

Spivak, Gayatri Chakravoty. 1989. "Who Claims Alterity?" In *Remaking History,* edited by Barbara Kruger and Phil Mariani, 269–92. San Francisco: Bay Press.

———. 1994. "Can the Subaltern Speak?" In *Colonial Discourses and Post-Colonial Theory: A Reader,* edited by Patrick Williams and Laura Chrisman, 66–112. Hertfordshire: Harvester Wheatsheaf.

Staeheli, Lynn A., and Victoria A. Lawson. 1994. "A Discussion of 'Women in the Field': The Politics of Feminist Fieldwork." *Professional Geographer* 46, no. 1:96–102.

Stanley, Liz. 1992. *The Auto/Biographical I/Eye*. Manchester, U.K.: Univ. of Manchester Press.

Stanley, Liz, and Sue Wise. 1983. *Breaking Out: Feminist Consciousness and Feminist Research*. London: Routledged and Kegan Paul.

Stanton, Domna, ed. 1984. *The Female Autograph: Theory and Practice of Autobiography from the Tenth to the Twentieth Century*. Chicago: Univ. of Chicago Press.

Stent, Gunther S. 1969. *The Coming of the Golden Age*. Garden City: Natural History Press.

Stinchcombe, Arthur L. 1968. *Constructing Social Theories*. Chicago: Univ. Chicago Press.

Superville, Philippa. 1993. " 'What the Hell Are You Doing Here?' Or Where Do I Fit into the IBG?" *Women and Geography Study Group Newsletter* 2:4–7.

Susser, Mervyn. 1988. *Epidemiology, Health and Society*. New York: Oxford Univ. Press.

Taussig, Michael. 1987. *Shamanism, Colonialism and the Wild Man: A Study in Terror and Healing*. Chicago: Univ. of Chicago Press.

Taylor, Charles. 1985. *Human Agency and Language*, Cambridge: Cambridge Univ. Press.

Taylor, S. Martin. 1984. "A Path Model of Aircraft Noise Annoyance." *Journal of Sound and Vibration* 96:243–60.

Taylor, S. Martin, Doug Sider, Christine Hampson, Kathi Wilson, and John Eyles. 1997. "Community Health Effects of a Petroleum Refinery." *Ecosystem Health* 3:27–43.

Therborn, Goran. 1980. *The Power of Ideiology and the Ideology of Power*. London: Verso.

Thompson, Becky, and Sanjeeta Tyagi, eds. 1996. *Names We Call Home: Autobiography on Racial Identity*. London: Routledge.

Thompson, E. P. 1968. *The Making of the English Working Class*. Harmondsworth, Eng.: Penguin.

Thompson, John B. 1983. "Rationality and Social Rationalization: An Assessment of Habermas' Theory of Communicative Action." *Sociology* 17:278–94.

Thrift, Nigel. 1983. "On the Determination of Social Action in Space and Time." *Environment and Planning D: Society and Space* 1:23–57.

Titmuss, Richard. 1962. *Income Distribution and Social Change*. London: Allen and Unwin.

———. 1965. *Commitment to Welfare*. London: Allen and Unwin.

Townsend, Peter, and Nick Davidson. 1982. *Inequalities in Health*. Harmondsworth, Eng.: Penguin.

Tuan, Yi-Fu. 1998. "A Life of Learning (1998 Charles Homer Haskins Lecture)." Occasional Paper no. 42. New York: American Council of Learned Societies.

Tuominin, Oiva. 1988. "Coincidence and Continuity in a Geographer's Life." In *Geographers of Norden: Reflections on Career Experiences,* edited by Anne Buttimer and Torsten Hägerstrand, 130–146. Lund: Lund Univ. Press.

United Nations Development Program. 1996. *Human Development Report.* New York: Oxford Univ. Press.

Unwin, Tim. 1992. *The Place of Geography.* London: Longman.

Valentine, Gill. 1998. " 'Sticks and Stones May Break My Bones': A Personal Geography of Harassment." *Antipode* 30, no. 4:305–32.

Visweswaran, Kamala. 1994a. *Fictions of Feminist Ethnography.* Minneapolis, Minn.: Univ. of Minnesota Press.

———. 1994b. "Sari Stories." In *Fictions of Feminist Ethnography,* 166–77. Minneapolis, Minn.: Univ. of Minnesota Press.

Wake, Marvalee H. 1993. "Two-Career Couples—Attitudes and Opportunites." *Bioscience* 43:238–40.

Walters, Vivienne, Susan French, John Eyles, Rhonda Lenton, K. Bruce Newbold, and Janet Mayr. 1997. "The Effects of Paid and Unpaid Work on Nurses' Well-being." *Sociology of Health and Illness* 19:328–47.

Weightman, Barbara. 1981. "Commentary: Towards a Geography of the Gay Community." *Journal of Cultural Geography* 1:106–12.

Western, John. 1981. *Outcast Cape Town.* London: Allen and Unwin.

White, Stephen K. 1988. *The Recent Work of Jürgen Habermas: Reason, Justice and Modernity.* Cambridge: Cambridge Univ. Press.

Wilson, Elizabeth. 1991. *The Sphinx in the City: Urban Life, the Control of Disorder, and Women.* Berkeley and Los Angeles: Univ. of California Press.

Wise, Sue. 1984. "Sexing Elvis." *Women's Studies International Forum* 7, no. 1:13–17.

Wolff, Janet. 1995. *Resident Alien: Feminist Cultural Criticism.* London: Polity Press.

Women and Geography Study Group. 1997. *Feminist Geographies: Explorations in Diversity and Difference.* Harlow, Essex: Longman.

Wood, Mary Elene. 1994. *The Writing on the Wall: Women's Autobiography and the Asylum.* Urbana: Univ. of Illinois Press.

Yaeger, Patricia. 1996. *The Geography of Identity.* Ann Arbor: Univ. of Michigan Press.

Young, Iris. 1990. *Justice and the Politics of Difference.* Princeton, N.J.: Princeton Univ. Press.

Zaretsky, Eli. 1976. *Capitalism, the Family and Personal Life.* New York: Harper Colophon.

Zelinsky, Wilbur. 1973a. "The Strange Case of the Missing Female Geographer." *Professional Geographer* 25, no. 2:101–105.

————. 1973b. "Women in Geography: A Brief Factual Account." *Professional Geographer* 25, no. 2:151–65.

Zweig, Ferdynand. 1948. *Men in the Pits.* London: Victor Gollancz.

Index